Computer Science Foundations and Applied Logic

Computer Science Foundations and Applied Logic is a growing series that focuses on the foundations of computing and their interaction with applied logic, including how science overall is driven by this. Thus, applications of computer science to mathematical logic, as well as applications of mathematical logic to computer science, will yield many topics of interest. Among other areas, it will cover combinations of logical reasoning and machine learning as applied to AI, mathematics, physics, as well as other areas of science and engineering. The series (previously known as *Progress in Computer Science and Applied Logic, https://www.springer.com/series/4814*) welcomes proposals for research monographs, textbooks and polished lectures, and professional text/references. The scientific content will be of strong interest to both computer scientists and logicians.

Luis Enrique Sucar

Causal Discovery

Foundations, Algorithms and Applications

 Birkhäuser

Luis Enrique Sucar 🆔
Department of Computer Science
INAOE
Tonantzintla, Puebla, Mexico

ISSN 2731-5754   ISSN 2731-5762 (electronic)
Computer Science Foundations and Applied Logic
ISBN 978-3-031-98344-3   ISBN 978-3-031-98345-0 (eBook)
https://doi.org/10.1007/978-3-031-98345-0

This book is published under the imprint Birkhäuser, www.birkhauser-science.com by the registered
company Springer Nature Switzerland AG
The registered company address is: Gewerbestrasse 11, 6330 Cham, Switzerland

If disposing of this product, please recycle the paper.

To my beloved wife, Doris, whose love and support made this journey possible; and to my wonderful children, Edgar and Diana, whose presence fills my life with inspiration and joy

Preface

Overview

Causal reasoning is a constant element in our lives as it is in human nature to constantly ask why. Acting in the world is conceived by human beings as intervening the world, and humans are able to learn and use causal relations. However, the development of causality as a scientific discipline is relatively recent. Causal models and causal reasoning had undergone an important development since the 90s of the previous century, which has been defined as a *causal revolution.* Why? Because causal models have several advantage with respect to associative techniques (such as statistical models and most of those used in machine learning). First, a causal model provides a deeper understanding of a domain, by identifying the direct and indirect causes of certain events, which is not necessarily true for associative models. Secondly, with causal models we can perform other types of reasoning that are not possible, at least in a direct way, with associative models. These other inference situations are: (i) *interventions,* in which we want to find the effects of setting a certain variable to a specific value by an external agent (note that this is different from observing a variable); and (ii) *counterfactuals,* where we want to reason about what would have happened if certain information had been different from what actually happened. Causality is also important for explanatory purposes, since an effect can be explained by its causes, regardless of the correlations it may have with other variables.

However, a challenge is how to learn causal models. Traditionally, to build a causal model we need to perform interventions in the *real* world, such as those in randomized clinical trails. Performing this type of interventions is costly, and in some cases unethical or impossible. But we have a lot of *observational* data, that is data obtained from natural processes, without an explicit intervention. This is a challenge, as only from observational data we cannot recover, in general, a unique causal model. There has been an increasing interest of the research community in learning causal models from observational data, what is known as *causal discovery,* with important developments in the last few years.

The objective of this book is to present an overview of the main algorithms and techniques for causal discovery, complementing other books on causality which mainly focus on causal reasoning. As this is a developing area, we describe in detail the basic classes of algorithms, and provide references to other variations and more recent developments. Although the focus is causal discovery, it provides the necessary background on causal modeling, with emphasis on causal graphical models; as well as causal reasoning, including interventions and counterfactuals. Some of the key features of the book are:

- It includes the necessary background material: a review of probability and graph theory, Bayesian networks, causal graphical models and causal reasoning.
- The main types of causal discovery are covered: learning from observational data, learning from interventional data, and learning from temporal data.
- It illustrates the application of causal discovery in practical problems, an important feature for students and practitioners.
- It includes some of the latest developments in the field, such as continuos optimization, causal event networks, causal discovery under subsampling, subject specific causal models, causal reinforcement learning, etc.
- Each chapter has a set of exercises, including suggestions for research and programming projects.

As causal discovery is a relatively new field, its applications are just developing. To illustrate the potential of causal modeling and discovery in many fields, and to complement the theory with practical applications; the book includes many examples of the application of causal discovery in a wide range of domains, such as:

- Eliciting brain effective connectivity.
- Causal discovery of the risk factors for COVID.
- Learning causal models of gene regulatory networks.
- Understanding the factors that affect decisions for the energy transition.
- Predicting the stock market.
- Causal inference in legal texts.
- Causal discovery from images.
- Drone navigation incorporating causal models.

Audience

This book can be used as a text book for an advanced undergraduate or a graduate course on causal discovery for students of computer science, engineering, social sciences, etc. It can also be used as a complement to a course con causality, together with another text on causal reasoning. It could also serve as a reference book for professionals that want to apply causal models in different areas, or anyone who is interested in knowing the basis of these techniques.

It does not have specific prerequisites, although some background in probability and statistics is recommended. It is assumed that the reader has a basic knowledge of mathematics at the high school level, as well as a certain background in comput-ing and programming. The programming exercises require some knowledge and experience with any programming language, such as R, Python, Matlab, etc.

Exercises

Each chapter (except the introduction and the application chapters) includes a set of exercises. These exercises are questions and problems designed to reinforce the understanding of the concepts and techniques presented in the chapter. There are also a few suggestions for research or programming projects (marked with "***"), which could be used as projects for a course.

Software

A list of software tools for causal discovery is provided in the Appendix.

Organization

The book is divided in three parts. Part I provides a general introduction and motivation for causal discovery, and reviews the required background on causal graphical models and causal reasoning.

Part II constitutes the core of the book, describing the main algorithms and techniques for causal discovery, divided in four chapters: (a) causal discovery from observational data, (b) causal discovery from interventional data, (c) causal discovery from temporal data, and (d) causal reinforcement learning.

Part III provides several examples of causal discovery in practice. It includes several applications in biomedicine and social sciences, as these are the main fields where causal modeling can have a greater impact as it is possible to make interventions. It concludes with a chapter about the interrelations between causal discovery and other areas of artificial intelligence, in particular machine learning and robotics.

Santa Cruz Tenerife, Islas Canarias,   Luis Enrique Sucar
Spain
April 2025

Acknowledgements This book grew out of the research of my group on causal discovery in the last few years. I am in debt to my collaborators and students which have contributed to some of the novel techniques and applications included in the book; as well as to the research community on causality at large, that have developed the foundations of this fascinating field.

I thank my collaborators, the common research projects and technical discussions have enriched my knowledge, and helped me write this manuscript. In particular I would like to mention my colleagues and friends: Armando Aguayo-Mendoza, David Danks, Adnan Darwiche, Hugo Jair Escalante-Balderas, José Ernesto Gómez-Balderas, José Martínez-Carranza, Manuel Montes y Gómez, Roberto Ley-Borrás, Samuel Montero-Hernández, Eduardo Morales-Manzanares, Rafael Murrieta-Cid, Felipe Orihuela-Espina, and Luis A. Pineda-Cortés.

I acknowledge those students with whom I have collaborated for their master or Ph.D. thesis. Some of the novel developments and application examples originated from their work. I thank them all, and will mention those whose work was more influential for this manuscript: Sergio Arredondo-Serrano, Sebastián Bejos-Mendoza, Mario de los Santos-Hernández, Ivan Feliciano-Avelino, Mauricio González-Soto, Armando Martínez-Ruiz, Arquímides Méndez-Molina, Samuel Montero-Hernández, Miranda Ramírez-Cruz, Verónica Rodríguez-López, and Nilda Xolo-Tlapanco. I appreciate the comments that Mario de los Santos has made of some earlier versions of this book.

Special thanks to my former student Sergio Arredondo-Serrano, who has revised the book, helping me to improve the writing, providing useful suggestions, and drawing many of the figures. Also my special gratitude to Julio Muñoz-Benítez, who has been a post-doctoral researcher during the writing of this book, helping me consolidate our research group, organizing our seminars and two international workshops on causal discovery, and sharing enriching discussions and ideas.

I would like to acknowledge the support from my institution, Instituto Nacional de Astrofísica, Óptica y Electrónica (INAOE), Puebla, Mexico, which is an excellent place for doing research and teaching, and has given me all the facilities to dedicate part of my time to write this book. Also, I thank the Instituto de Astrofísica de Canarias, Tenerife, Spain, that received me for a sabbatical year, providing an ideal environment to conclude this book.

Last, but certainly not least, I extend my deepest gratitude to my family. To my parents, Fuhed[†] and Aida, who instilled in me a passion for learning, a strong work ethic, and unwavering support throughout my studies. In particular, to my father, whose remarkable books not only inspired me but perhaps also passed me the genes for writing. To my brother Ricardo[†], and my sisters, Shafía and Beatriz, whose encouragement and belief in my dreams have always been a source of strength. And most importantly, to my beloved wife, Doris, and my children, Edgar and Diana—your love and support are the foundation upon which I stand, the driving force that keeps me going.

Competing Interests The author has no competing interests to declare that are relevant to the content of this manuscript.

Contents

Acronyms

AI	Artificial Intelligence
ATE	Average Treatment Effect
AUC	Area Under the Curve
BCCD	Bayesian Constraint-based Causal Discovery
BDeU	Bayesian Dirichlet equivalent and Uniform
BIC	Bayesian Information Criterion
BN	Bayesian Network
BS	Bayesian Score
CBN	Causal Bayesian Network
CD	Causal Discovery
CGNN	Causal Generative Neural Network
CMI	Conditional Mutual Information
CPDAG	Completed Partially Directed Acyclic Graph
CPT	Conditional Probability Table
CRL	Causal Reinforcement Learning
DAG	Directed Acyclic Graph
DBN	Dynamic Bayesian Network
DQN	Deep Q-Learning
DRL	Deep Reinforcement Learning
DT	Decision Tree
EM	Expectation-Maximization
FASK	Fast Adjacency Skewness
FCI	Fast Causal Inference
FGES	Fast GES
fMRI	functional Magnetic Resonance Imaging
fNIRS	functional Near-Infrared Spectroscopy
FPR	False Positive Rate
GAN	Generative Adversarial Network
GCI	Graph-based Causal Inference
GES	Greedy Equivalence Search
GFCI	Greedy and Fast Causal Inference
GLM	Gaussian Linear Model

GNN	Graph Neural Network
ICA	Independent Component Analysis
ID	Influence Diagram
KB	Knowledge Base
KG	Knowledge Graph
LiNGAM	Linear, Non-Gaussian, Acyclic Model
MAG	Maximal Ancestral Graph
MAP	Maximim a Posteriori Probability
MB	Markov Blanket
MCI	Momentary Conditional Independence
MDL	Minimum Description Length
MDP	Markov Decision Process
MEC	Markov Equivalence Class
MMD	Maximum Mean Discrepancy
MPP	Multivariate Point Processes
PAG	Parcial Ancestral Graph
PCMCI	PC with Momentary Conditional Independence
PGM	Probabilistic Graphical Model
PN	Probability of Necessity
PNS	Probability of Necessity and Sufficiency
PS	Probability of Sufficiency
RL	Reinforcement Learning
SCM	Structural Causal Model
SEM	Structural Equations Model
SHD	Structural Hamming Distance
TEN	Temporal Event Network
TPR	True Positive Rate
TTM	Transtheoretical Model
UAV	Unmanned Aerial Vehicle

Notation

T	True
F	False
$A, B, C, ...$	Propositions (binary variables)
$\neg A$	Not A (negation)
$A \wedge B$	A and B (conjunction)
$A \vee B$	A or B (disjunction)
$A \in B$	A is an element of B
$\forall(X)$	Universal quantifier: for all X
$\exists(X)$	Existential quantifier: exists an X
$C \cup D$	Union of two sets
$C \cap D$	Intersection of two sets
$n!$	Factorial of n, $n! = n \times (n-1) \times (n-2) \times ...1$
$\binom{n}{r}$	Combinations of r from n, $\binom{n}{r} = \frac{n!}{r!(n-r)!}$
$exp(x)$	Exponential of x, $exp(x) = e^x$
$log_n(x)$	Logarithm base n of x
$|\mathbf{X}|$	The dimension or number of states of a discrete variable or vector $\mathbf{X}$
$\sum_i X_i$	Summation over X
$\prod_i X_i$	Product over X
$\int_a^b f(x)dx$	Integral of $f(x)$ from a to b
Ω	Sample espace
μ	Mean
σ^2	Variance
σ_{XY}^2	Covariance of XY
σ	Standard deviation
$N(\mu, \sigma^2)$	Normal distribution with mean μ and standard deviation σ
X	A variable
x	A particular value of a variable, $X = x$
$\mathbf{X}$	A vector of variables, $\mathbf{X} = X_1, X_2, ..., X_N$
$\mathbf{x}$	A particular realization of vector $\mathbf{X}$, $\mathbf{x} = x_1, x_2, ..., x_N$

$\mathbf{X}_{1:T}$	Vector of variable X from $t = 1$ to $t = T$, $X_{1:T} = X_1, X_2, ..., X_T$
$I(m)$	Information
$H(M)$	Entropy
$E(X)$	Expected value of a random variable X
$ArgMax_x F(X)$	The value of X for which the function F is maximum
$P(X = x)$	Probability of X being in state x; for short $P(x)$
$P(\mathbf{X} = \mathbf{x})$	Probability of $\mathbf{X}$ being in state $\mathbf{x}$; for short $P(\mathbf{x})$
$P(x, y)$	Probability of x and y
$P(x \vee y)$	Probability of x or y
$P(x \mid y)$	Conditional probability of x given y
$P(x) \sim y$	The probability of x is *proportional* to y, that is $P(x) = k \times y$
$\mathbf{P}(X)$	Cumulative distribution function of a discrete variable X
$P(X)$	Probability function of a discrete variable X
$F(X)$	Cumulative distribution function of a continuous variable X
$f(X)$	Probability density function of a continuous variable X
$I(X, Y, Z)$	X independent of Z given Y
$P(X = x \mid do(Y = y))$	Probability of $X = x$ given the intervention $Y = y$
$P_{do(Y=y)}(x) = P_y(x)$	Probability of $X = x$ given the intervention $Y = y$
$P_{\mathbf{y}}(\mathbf{X})$	Probability distribution of $\mathbf{X}$ given the intervention $\mathbf{Y} = \mathbf{y}$
$G(V, E)$	Graph G with set of vertices V and set of edges E
$E(V_j, V_k)$	Edge E between vertices V_j and V_k in a graph
$Adj(V)$	Vertices adjacent to vertex V in a graph
$Pa(X)$	Parents of node X in a directed graph
$A \to B \to C$	Sequential arcs
$A \leftarrow B \to C$	Divergent arcs
$A \to B \leftarrow C$	Convergent arcs (V structure)
$A \to B$	A causes B
$A \leftrightarrow B$	There is a cofactor that causes A and B
$A \leftarrow\!\!\!\circ\, B$	B causes A or a cofactor that causes A and B
$\mathbf{I}$	Identity matrix
$\mathbf{A}^{-1}$	Inverse of matrix $\mathbf{A}$
$\mathbf{AB}$	Matrix multiplication
π	A policy for an MDP or POMDP: mapping from states to actions
V^{π}	Value function following policy π
γ	Discount factor

Part I

Foundations

The first chapters of the book include a general introduction to causality, and provide the foundations required for the rest of the book: causal graphical models and causal reasoning.

Introduction

1

Abstract

This chapter provides a general introduction to causality. It includes a brief history of causality, an overview of different causal modeling approaches, the definition of causality that will be considered in the rest of the book, and some application examples of causal discovery.

1.1 Introduction

Causal reasoning is a constant element in our lives as it is in human nature to constantly ask why [5,8]. Acting in the world is conceived by human beings as intervening the world, and humans are able to learn and use causal relations while making choices. Causality has to do with cause–effect relations; that is, identifying when there are two (or more) related phenomena, which is the *cause* and which is the *effect*. However, there could be a third explanation, namely, there is another phenomena which is the *common cause* of the original two phenomena of interest.

L. E. Sucar, *Causal Discovery*, Computer Science Foundations and Applied Logic,
https://doi.org/10.1007/978-3-031-98345-0_1

Fig. 1.1 An example of a causal graph

Probabilistic models do not necessarily represent causal relations. For instance, consider the following two Bayesian networks[1]: BN1: $A \rightarrow B \rightarrow C$ and BN2: $A \leftarrow B \leftarrow C$. From a probabilistic perspective, both represent the same set of dependency and independency relations: direct dependencies between A and B, B and C; and $I(A, B, C)$, that is, A and C are independent given B. However, if we define that a directed link $A \rightarrow B$ means A causes B, both models represent very different causal relations.

Figure 1.1 depicts an example of a causal graphical model, which represents the causal knowledge that rain makes the street wet, and that a wet street is slippery. From a traditional statistical point of view, the model represents that *Rain* is independent of *Slippery* given *Wet-street*. If we reverse the arcs in Fig. 1.1 this independence relation still holds; however, causal models can express that rain is the actual cause of the street being wet, and not the other way around (which common sense helps us determine that it makes no sense).

Given that we can model model complex phenomena and solve problems/tasks associated them with different techniques, such as probabilistic graphical models, and also that causality is for many a complex and controversial concept, we may ask ourselves, why do we need causal models? There are several advantages to causal models, in particular, graphical causal models.

First, a causal model provides a deeper understanding of a domain, by identifying the direct and indirect causes of certain events, which is not necessarily true for associative models such as Bayesian networks. Secondly, with causal models we can perform other types of reasoning that are not possible, at least in a direct way, with associative models. These other inference situations are: (i) *interventions*, in which we want to find the effects of setting a certain variable to a specific value by an external agent (note that this is different from observing a variable); and (ii) *counterfactuals*, where we want to reason about what would have happened if certain information had been different from what actually happened. We will discuss these two situations in more detail in the rest of the book. Causality is also important for explanatory purposes, since an effect can be explained by its causes, regardless of the correlations it may have with other variables.

Causal reasoning is also useful for solving the *credit assignment problem*; that is, choosing the right event that predicts (gives credit) to other important event. There could be many events and actions that occur before an important event, by identifying the cause of this important event we can predict its occurrence in the future. For instance, if we can identify the risk factors (causes) of certain disease, we can avoid them in the future.

---

[1] Bayesian networks represent the dependence and independence relations between a set of variables as a directed acyclic graph; they will be covered in detail in Chap. 2.

We need to study causation to understand how and why causes influence their effects. For this we requiere: (i) a formal definition of causation, (ii) causal models to represent the causal knowledge and assumptions, (iii) a method to derive conclusions from causal models and data, (iv) a way to build or learn the causal models.

Before we address these aspects, we will present a brief history of the development of causality.

1.2 A Brief History

Figure 1.2 shows some of the main developments in the history of causality, starting from the Greek philosophers to the present time. In the next sections we will cover these and other important contributions to causality, from the philosophical fundamentals to the mathematical modeling.[2]

1.2.1 Philosophical Foundations

As in many other sciences, the study of causation starts with the ancient Greeks. Greek philosophers used the noun *aitia*, translated as 'cause' (or 'reason' or 'explanation'), to mean responsibility or give credit. To assign to X the status of *aitia* to Y, means that to understand or make sense of Y, we should look for how it is related to X. Given the various meanings of *aitia*, for ancient Greeks the "causes of things" were not only about the identity of the causes, but also about explanations. Several Greek philosophers discussed about explanations, but Aristotle gave them the most

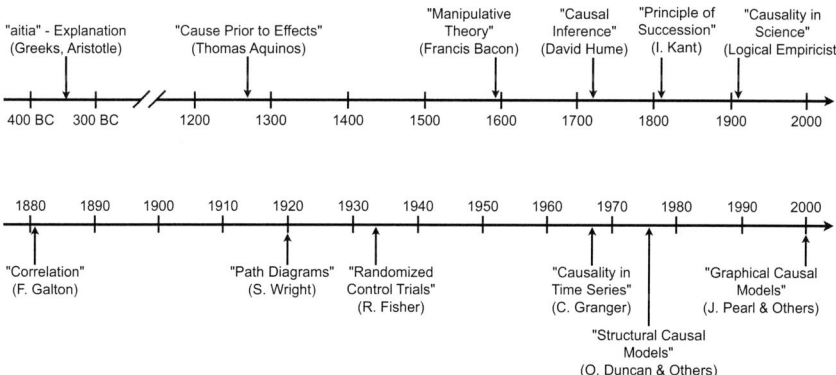

Fig. 1.2 A timeline highlighting some of the main developments in the history of causality

---

[2] This section is based on [1,5]. The first part focuses on philosophical aspects, and the second has to do more with modeling aspects.

important philosophical development. "Knowledge is the object of our inquiry, and we do not think we know a thing till we have grasped the 'why' of it (which is to grasp its primary cause)" (Aristotle, Physics 2.3) [1].

Medieval philosophers broadly had a similar notion of causality as now: causes are prior to their effects, which they produce and the existence of which they explain. However, analyzed more closely, medieval notions of causality are different from contemporary ones. According to Thomas Aquinas (*Summa Theologiae*), since every cause is prior to its effect, none can be its own cause, because it would be prior to itself which is impossible. It is not possible to proceed to infinity, so without a first cause there would be no other causes and effects. This first cause is what everyone calls God! Aquinos distinguishes between essential (*per se*) and accidental causation. He argues that it is possible to proceed to infinity by accidents in agents causes, but it is impossible for essential causes.

Early moderns confronted an abundance of causal explanations; they inherited the doctrine of the four causes from Aristotle: material, formal, efficient, and final cause. Material cause relates to the physical properties of something. Formal cause is the structure of design of a being. The efficient cause is the moving cause of a change or movement. The final cause is the change or movement for the sake of a thing to be what it is. For example, for a chair the material cause could be wood, the formal cause the design of the chair, the efficient cause is the carpenter who make the chair, and the final cause is that the chair is for someone to sit.

Several scientists and philosophers contributed to the development of causality during this period. Bacon proposes a *manipulative* theory of causation: "If the cause is not known the effect cannot be reproduced". Descartes supports the principle of *efficient* causation: "There must be reality in the efficient and total cause as in the effect of that cause". He suggests that the effect must be *like* its cause. Hobbes has a materialistic point of view, cause and effect are identified with the agent (cause) and the patient (effect); the actor and the acted upon. He thinks that causation must be instantaneous. Spinoza and Leibniz embrace a *explanatory rationalism*, which provides a satisfactory answer to any "Why" question. They consider that systems are deterministic –everything is caused—and everything is necessitated—causes necessarily bring about their effects.

David Hume is well known for having a theory of causation. He asserts that we should first understand *causal inference*, which is the discovery of causal relations. Following this perspective, Hume gives two definitions of "cause": (i) "An object precedent and contiguos to another, and where all objects resembling the former are placed in like relations of precedency and contiguity to those objects that resemble the latter" (external conditions); (ii) "An object precedent and contiguos to another, and so united with it that the idea of the one determines the mind to form the idea of the other, and the impression of the one to form a more lively idea of the other" (internal conditions). Causal relations are discovered by the process of causal inference, so objects of a type that are constantly conjoined produce association, and an impression of determination (the occurrence of one type of object determines the occurrence of the other) or necessary connection.

Kant was interested in causality, he views causal relations as what binds substances together to form a single world. Kant proposes two causal principles, the Principle of Succession and the Principle of Coexistence; establishing that causality and mutual interaction are required for succession and coexistence. His concept of causality is related to that of an *event*, defined as a change of an object state; so what ever determines the succession of states of an object is what we mean by cause. Kant also explored how causality is applied to the sciences, in particular physics and biology.

In the *Logical Empiricist*, Frank, Schlick and Reichenbach, studied causality's connection to science, in particular physics; they sought to make the theory of causation applicable to scientific progress. There existed a close connection between their views on causality and probability.

The law of causality: "If, in the course of time, a state of the universe A is followed by the state B, then whenever A occurs B will follow", can neither be confirmed or disproved by experience, affirms Frank. On the other hand, our whole science and even our practical life is apparently based on the law of causality—each *manipulation* is accompanied by the expected result.

Schlick holds that the principle of causality is "A general expression of the fact that everything which happens in nature is subject without exception to valid laws. The causal determinacy of the world extends only one dimension, the direction of time". However, this was questioned by the developments in quantum theory, "If the new results in quantum theory were true, the world, in the last resort, would be handed over to chance". Later, he modified his view with a kind of compromise: "The contribution made by the present physics to the causality problem does not contest the validity of the principle of causality ... the novelty consists in the discovery that the laws of nature set a limit of principle to the exactness of prediction".

A basic idea developed by Reichenbach, is that for the principle of causality to be applicable to the description of physical phenomena it must be supplemented by a second principle—the principle of the lawful distribution or the principle of the continuous probability function. Causality guaranteed that there existed a functional relation, while the second principle implied that the disturbing factors remained small enough. Given this second principle, all physical (causal) laws become probabilistic. Reichenbach's inductive principle of causality: starting from a presumed law we find further relevant causes that lead to a modified function, by iterating with observations we can arrive to two different conclusiones, (I) we have found a causal law, or (II) the connection between the observations is random.

1.2.2   Causal Models

Causality was approached by some of the creators of statistics, however for several reasons it was abandoned.

Galton was looking for a causal explanation of why descendants of genius do not tend to be geniuses, or sons of tall people do not tend to be taller—regression to the mean. Although he initially was interested in causation, he was not able to find a causal explanation—fathers and sons of tall people tend to be shorter—he divorced statistics from causation. He analyzed how different quantities are related and discovered the concept of *correlation*, the field of statistics was started.

Pearson was Galton's disciple and probably the most famous statistician in history. He completed the task of separating causation from statistics, in Pearson's words [5]: "I interpreted ... Galton to mean that there was a category broader than causation, namely correlation, of which causation was only the limit, and that this new conception of correlation brought psychology, anthropology, medicine and sociology in large part into the field of mathematical treatment. It was Galton who first freed me from the prejudice that sound mathematics could only be applied to natural phenomena under the category of causation." Pearson found that when he analyzed the relation between the length and the breadth of women's and men's skulls separately the correlation was negligible, but when analyzed together there is a significant correlation! He used this example as another argument against causality (although it can be explained by a causal model—the relation has to do with the difference between men's and women's skulls). Pearson's character and strong position make any mention of causality seem almost as a sin!

Many years later, the first graphical representation of causality was proposed by Sewall Wright [9]. Wright studied genetics and analyzed the coat color of Guinea pigs. He found that the changes in color could not be explained by genetics alone and hypothesized that it was a combination of genes and environmental factors. He developed a model that explains the factors that determine the coat color of Guinea pigs, he called them *path diagrams* –the first causal models! (published 1920). An example of a path diagram is depicted in Fig. 1.3.

The arrows in a path diagram all correspond to causal relations, whose strength is quantified by the associated parameters (small letters), called path coefficients. Wright showed that if we know the casual model (path diagram) then we can predict the correlations in the data. It was the first "bridge" between causation and probability. It is also possible to go the other way: from the correlations measured in the data to the hidden causal quantities, by solving algebraic equations.

Statisticians criticize Wright's work. "Causation is just perfect correlation" (not necessarily, it increases the probability, like in the case of smoking causes cancer). "Representing scientific knowledge as a diagram is a basic fallacy" (now it is the way that scientists represent causal knowledge). Path analysis was basically ignored for 40 years! (from 1920 to 1960).

In the second half of the twentieth century there was a rebirth of causal models, including *structural equation models* in sociology and *simultaneous equations models* in economy. Rubin [7] introduced the *potential outcome framework*, which contributed to the estimation of causal effects from observational data. Finally, in the 1990's, causal graphical models emerged.

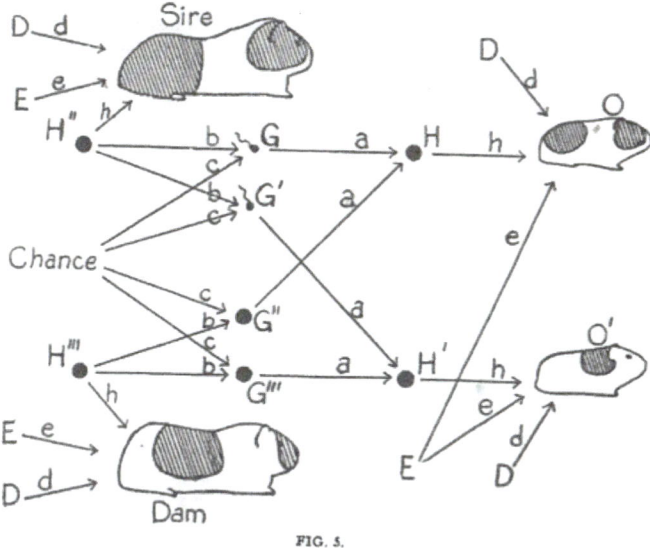

FIG. 5.

Fig. 1.3 An example of a path diagram. The capital letters represent the different factors: D, development; E, environmental factors; G, genetic factors; H, combined parents' factors. The arrows represent causal relations quantified by the corresponding parameters (in small letters). Figure from [9]

1.3   Causal Inference

Causal inference in terms of the effect of certain drugs and treatments has been studied for a long time in medicine, and statistical procedures have been developed to estimate the *causal effect*, for instance, of a novel drug. Next we give a brief review of these developments.[3]

Consider that a novel treatment for certain disease has been developed, and when applied to a person she dies; so the question is if the treatment *caused* the death? Or maybe there are other aspects already present in this person that were the causes of her sudden death. That is, we want to determine the *causal effect* of this treatment. For this it will be necessary to know what would had happen if the person was not given the treatment (control), will she have died or will still be alive? A causal effect is then a comparison of the outcome a person will have if given the treatment and the outcome the same person will have given the control. Causal inference is challenging because for any individual, we cannot see both outcomes [6] (which is known as the *fundamental problem of causal inference*).

A solution is to apply the treatment and control conditions to a group of persons, and then compare the results. For this a critical aspect is how to decide who will

---

[3] For a more detailed analysis see [6].

receive the treatment and who will receive the control. This should be done randomly; that is, for each participant throw a coin, and depending on the outcome assign her to the control or treatment group. This is known as a *randomized control trail*, and is the standard for medical studies, comparing, for instance, different drugs to treat a disease.

A random selection is critical as there could be other variables that affect the results, such as the age or sex of the individual, these are known as *cofactors*. A random assignment to the treatment and control groups will make that these cofactors have no effect on the results, as the average of the different cofactors will tend to be similar for both groups. For this, both groups should have been selected randomly, and the larger the better (given the law of large numbers in statistics).

As a result of a randomized control trail we can estime the *Average Treatment Effect* (ATE) as the difference between the results of both groups. For example, consider we want to measure the effect of a new vaccine, the treatment group receives the vaccine and the control group a placebo. The result is measured in terms of the survival after one month that the procedure, vaccine or placebo, was applied. Thus, we can measure the effect of the vaccine as the difference between the survival rate of both groups:

$$ATE = r_T - r_C, \tag{1.1}$$

where r_T is the proportion of individuals that survived given the treatment and r_C is the proportion of individuals that survived given the control. If ATE is positive, the vaccine has a causal effect in preventing the disease, if negative it has an opposite effect, and if near zero it has no effect (a more detailed statistical analysis could be done to analyze the results that depends on the size of the sample).

Performing a randomized control trail is complex, expensive and in some cases unethical. Evaluating if smoking caused cancer remained controversial for a long time as it was unethical to force a group of people to smoke and compare them with other group that did not smoke. In this case, as in many other, we have *observational data*, obtained from the natural phenomena without a control trail. In the smoking example, we can obtain data of people that smoke and did not smoke, and if they later developed cancer. So the question is, can we estimate a causal effect given observational data?

A problem with observational data is that individuals in the different groups are not chosen randomly, i.e., a person decides or not to smoke based on multiple factors. As a result, the treatment and control groups can be very different in terms of several cofactors, such as sex, age, income level, etc. So a difference in the results between both groups could be due to one or more of these cofactors and not to the treatment. In the controversy about smoking and cancer, a possible explanation was that there was a gen that made people smoke and also increased the probability of developing cancer, a cofactor.

A solution to the problem is to try to match for covariates, so both groups have similar proportions for all the covariates. That is, select a subgroup from both groups, such that both subgroups are similar in terms of sex, age, income level, eduction, etc.; for all the measured variables. This should be done for the combination of values for all the cofactors, calculating what is known as the *propensity score*. That is, the probability of the variable of interest (smoking) given each combination of values of the cofactors should have a similar probability distribution for both subgroups. By matching the propensity score, the measured covariates are balanced, and we can have certain confidence that the results obtained are unbiased.

There are several problems with this approach. One is that the treatment and control groups should be large enough so that there are sufficient individuals for each value of the cofactors, so that two matching subgroups can be obtained. But the more difficult problem is that of unmeasured cofactors. What if there is other cofactor that is discovered in the future that has important impact in the results?

The ideal solution to measure the causal effect of interventions is to perform a randomized control trial. An alternative is to have a "good" causal model of the domain of interest so that we can simulate the intervention in this model and estimate its effect on the variable of interest. Although it might not be possible to discover a causal model in every domain, for some it is feasible, and even an incomplete or partial model can be useful. How to discover a causal model from data is the main topic of this book.

1.4 Definition of Causality

There are several interpretations of causality, here we adopt the *manipulationist* interpretation [4,8]: "Manipulation of a cause will result in a manipulation of the effect". For example: *if we force a rooster to crow it will not cause the sun to rise, instead if the sun rises it causes the rooster to crow.*

We adopt the formal definition of causality provided by Spirtes et al. [8] which states that causation is a stochastic relation between events in a probability space; this is, some event (or events) causes another event to occurr. Let Ω, F, P^4 be a finite probability space, and consider a binary relation $\rightarrow \subseteq F \times F$ which is:

- Transitive: If $A \rightarrow B$ and $B \rightarrow C$, $\forall A, B, C \in F$, then $A \rightarrow C$.
- Irreflexive: For all $A \in F$ it doesn't hold that $A \rightarrow A$.
- Antisymmetric: For $A, B \in F$ such that $A \neq B$ if $A \rightarrow B$ then it does not hold that $B \rightarrow A$.

A is a cause of B if $A \rightarrow B$. It is important to note that an event may have more than one cause, and that not necessarily each one of these causes is sufficient to produce the effect.

---

[4] Where Ω is the sample space, F is s a set of subsets of Ω, and P is a probability function over Ω.

1.5 The Ladder of Causation

Pearl [5] introduces the *ladder of causation* that defines the different levels of reasoning: association, intervention and counterfactuals.

1. Association: Detecting regularities in the world: $P(X|Y)$. Example: Probability of heart attack given we that observe that the person drinks wine (correlation).
2. Intervention: Predicting the effects of deliberative actions, choosing the appropriate one to obtain certain effects: $P(X|Do(Y))$.[5] Example: Probability of heart attack given we administer aspirin (intervention).
3. Counterfactuals: Imagining, introspection, understanding: $y = y1 : P(X|Do(Y = y2)$. Example: A person died of a heart attack, would he have died if we did not administered aspirin? (what if).

   Traditional statistical and computational models can only reason at the associative level; for reasoning about interventions and counterfactuals we require causal models.

1.6 Different Approaches for Causal Modeling

Next we briefly introduce some types of causal modeling approaches, these are described in more detail in the following chapters.

1.6.1 Granger Causality

Granger proposed a practical definition of causality based on prediction improvement. *X causes Y* if there is some information in X relevant for Y that is not contained in Y's past as well as the past of "all the information in the Universe". In practice, typically only Y's past is used. Measuring prediction improvement can be operationalized in different ways. The most common framework are vector autoregressive models (VAR):

$$\mathbf{X}_t = \sum_{\tau=1}^{\tau_{max}} \Phi(\tau)\mathbf{X}_{t-\tau} + \eta_t \tag{1.2}$$

where $\mathbf{X}_t = (X_t^1, X_t^2, ..., X_t^N)$ is a time series that contains N variables, $\Phi(\tau)$ is the $N \times N$ coefficient matrix at lag τ, τ_{max} some maximum time lag, and η denotes an independent noise term. Here, X^i causes X^j if any of the coefficients at lag τ is non-zero; which represents a causal link $X^i \rightarrow X^j$ at lag τ.

---

[5] The $Do(X)$ operator is introduced to differentiate observing X from setting X through and intervention; this will be discussed in detail in Chap. 2.

1.6.2 Causal Bayesian Networks

A Causal Bayesian network (CBN) is a directed acyclic graph, G, in which each node represents a variable and the arcs in the graph represent *causal* relations; that is, the relation $A \rightarrow B$ represents some physical mechanism such that the value of B is directly affected by the value of A. This relation can be interpreted in terms of *interventions*—setting of the value of some variable or variables by an external agent. For example, assume that A stands for a water sprinkler (OFF/ON) and B represents if the grass is wet (FALSE/TRUE). If the grass (B) is originally not WET and the sprinkler is set to ON by an intervention, then B will change to TRUE.

Causal networks represent stronger assumptions than Bayesian networks, as all the relations implied by the network structure should correspond to causal relations. Thus, all the parent nodes, $Pa(X)$, of a certain variable, X, correspond to the direct causes of X. This means that if any of the parent variables of X, or any combination of them, is set to a certain value via an intervention, this will have an effect on X.

1.6.3 Functional Causal Models

A functional causal model consists of a set of *structural* equations of the following form:

$$X_i = f_i(Pa(X_i), U_i), i = 1, ..., n \qquad (1.3)$$

where $Pa(X_i)$ are the *parents* of X_i, that is the variables that directly determine the values of X_i; and U_i represent the error or disturbances associated to each equation (could be due to other factors not included in the model). Although these equations can take any form, it is common to restrict them to linear equations, and this type of models are known as linear *structural equation models* (SEMs).

In contrast with algebraic equations, structural equations change their meaning under algebraic operations; these preserve the solution of the equations but alter their causal meaning.

1.7 Applications

Causal discovery is being applied in different areas. On one hand, having a causal model of certain phenomena, can help to have a better and deeper understanding of these phenomena. On the other hand, a causal model can be used to predict the effects of different interventions without the need to make them in the "real world", which could be costly or unethical. These are some examples of applications of causal discovery:

- **Learning a Causal model for ADHD**: Analyzing the different factors causally related to the Attention-Deficit/Hyperactivity Disorder (ADHD) in children.
- **Decoding Brain Effective Connectivity**: Discovering the effective (causal) connectivity network in the brain from neuro-images.
- **Understanding COVID-19**: Finding the main factors that cause hospitalization and death for people affected with COVID-19.
- **Causal Discovery for Robotics Applications**: Discovering and applying causal models in several robotics tasks, such as visual navigation, manipulation, and human-robot interaction.
- **Building a Causal Model for the Energy Transition**: Elucidating the different factors that people consider for their decisions related to reducing energy consumption at home and for transportation.
- **Causal Reinforcement Learning for Industrial Settings**: Simultaneously learning a dynamic causal Bayesian network and a control policy for improving the performance of industrial processes.
- **Causal Learning for Legal Case Matching**: A causal learning framework for legal case matching respecting the corresponding law articles.
- **Learning Causal Models in Economics**: Understanding the causal relationship between minimum wage changes and the level of employment in a particular industry.
- **Causal Discovery for Marketing**: Learning causal structures for marketing mix models.
- **Causal Models in Computer Vision**: Finding causal relations between objects in an image.

These and other applications will be described in more detail in the third part of the book.

1.8 Overview of the Book

The book is divided into three parts.

Part I provides the foundations for the rest of the book. After this introduction, in Chap. 2 we introduce causal graphical models; including a brief review of the probability theory and Bayesian networks. Chapter 3 describes the main algorithms for causal reasoning, including predictions and counterfactuals.

Part II is the core of the book, covering the different algorithms for causal discovery. It is divided into four chapters: (i) causal discovery from observational data; (ii) causal discovery from interventional data; (iii) causal discovery from time series; and (iv) causal reinforcement learning, in which, after a brief introduction to reinforcement learning (RL), how causal discovery and RL are combined is described.

Part III presents several applications of causal discovery, including biology and medicine, social sciences, and robotics and artificial intelligence.

1.9 Additional Reading

A gentle introduction to causality and causal reasoning is presented in the book by Pearl and Mackenzie [5]. Two more technical books are [4,8], [4] is mainly about representation and reasoning, while [8] also includes learning. The Oxford Book on Causality [1] includes a history of causality and a description of several approaches for causal modeling. Rosenbaum [6] provides an excellent introduction to causal inference, in particular to randomized control trials.

Acknowledgements The section about the history of causality is based on [1,5]. The ladder of causality was originally introduced by [5]. The section on causal inference is inspired on [6]. Figure 1.3 is taken from [9] that is public domain. The images at the beginning of this, and the rest of the chapters, were generated with ChatGPT, in particular GPT-4-turbo variant.

References

1. Beebee H, Hitchcock C, Menzies P (eds) (2012) The Oxford handbook of causation. Oxford University Press, Oxford UK
2. Koller D, Friedman N (2009) Probabilistic graphical models: principles and techniques. MIT Press, Cambridge
3. Pearl J (1988) Probabilistic reasoning in intelligent systems: networks of plausible inference. Morgan Kaufmann, San Francisco
4. Pearl J (2009) Causality: models, reasoning and inference. Cambridge University Press, New York
5. Pearl J, Mackienze D (2018) The book of why. Basic Books
6. Rosenbaum PR (2023) Causal inference. The MIT Press, Cambridge, Massachussets
7. Rubin DB (1974) Estimating causal effects of treatments in randomized and nonrandomized studies. J Educ Psychol 66:688–701
8. Spirtes P, Glymour C, Scheines R (2000) Causation, prediction, and search. MIT Press
9. Wright S. (1920) The relative importance of heredity and environment in determining the piebald pattern guinea-pigs. In: Proceedings of the National Academy of Sciences, pp 320–332

Causal Graphical Models

2

Abstract

This chapter introduces causal graphical models. We start with a review of the required background on probability theory and graph theory. Then, we present a general overview of Bayesian networks, including representation and reasoning. Next, we describe causal Bayesian networks, as an extension of Bayesian networks for causal modeling and reasoning about interventions (2nd level in the ladder of causation). Finally, we introduce structural equation models, required for counterfactual reasoning (3rd level in the ladder of causation).

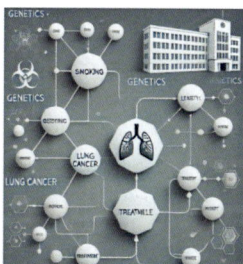

2.1 Introduction

Causal relations in general are not deterministic, they have some level of uncertainty. For instance, the advice that we find in all cigarets, "Smoking causes cancer", implies that if you smoke the probability that you will develop cancer increases, but not all persons that smoke get cancer. Thus, it is necessary to consider a way to include uncertainty within causal models. This could be modeled in different forms: causal Bayesian networks implicitly include uncertainty in the parameters of the model (the conditional probabilities of a variable given its parents); functional causal models include error terms in the equations.

Thus, before we go deeper into causal models, we provide a brief overview of probability theory. As these models are represented by graphs, we include a brief summary of graph theory. We also provide a review of Bayesian networks, as these are the basis of causal Bayesian networks. Then, we describe the two main types of causal models considered in this book: causal Bayesian networks and functional equation models. In this chapter we focus on representation, and in the next on inference.

2.2 Overview of Probability Theory

Probability theory was originated in games of chance and has a long and interesting history; it has developed into a mathematical language for quantifying uncertainty. We will consider the logical or normative approach and define probabilities in terms of the degree of plausibility of a certain proposition given the available evidence [4]. Based on Cox's work, Jaynes establishes some basic desiderata that this degree of plausibility must follow [4]:

- Representation by real numbers.
- Qualitative correspondence with common sense.
- Consistency.

Based on these intuitive principles, we can derive the axioms of probability theory:

1. $P(A)$ is a continuous function in $[0, 1]$.
2. $P(A, B \mid C) = P(A \mid C)P(B \mid A, C)$ (product rule).
3. $P(A \mid B) + P(\neg A \mid B) = 1$ (sum rule).

Where A, B, C are propositions (binary variables) and $P(A)$ is the probability of proposition A. $P(A \mid C)$ is the probability of A given that C is known, which is called *conditional probability*. $P(A, B \mid C)$ is the probability of A AND B given C (logical conjunction) and $P(\neg A \mid C)$ is the probability of NOT A (logical negation) given C. These rules are equivalent to the most commonly used Kolmogorov axioms. From these axioms, all conventional probability theory can be derived.

2.2.1 Basic Rules

The probability of the disjunction (logical sum) of two propositions is given by the *sum rule*: $P(A + B \mid C) = P(A \mid C) + P(B \mid C) - P(A, B \mid C)$; if propositions A and B are mutually exclusive given C, we can simplify it to: $P(A + B \mid C) = P(A \mid$

$C) + P(B \mid C)$. This can be generalized for N mutually exclusive propositions to:

$$P(A_1 + A_2 + \cdots A_N \mid C) = P(A_1 \mid C) + P(A_2 \mid C) + \cdots + P(A_N \mid C) \quad (2.1)$$

In the case that there are N mutually exclusive and exhaustive hypotheses, $H_1, H_2, \ldots, H_N$, and if the evidence B does not favor any of them, then according to the principle of indifference: $P(H_i \mid B) = 1/N$.

According to the logical interpretation there are no *absolute* probabilities, all are conditional on some background information.[1] $P(H \mid B)$ conditioned only on the background B is called a *prior* probability; once we incorporate some additional information D we call it a *posterior* probability, $P(H \mid D, B)$. From the product rule we obtain:

$$P(D, H \mid B) = P(D \mid H, B)P(H \mid B) = P(H \mid D, B)P(D \mid B) \quad (2.2)$$

From which we obtain:

$$P(H \mid D, B) = \frac{P(H \mid B)P(D \mid H, B)}{P(D \mid B)} \quad (2.3)$$

This last equation is known as the *Bayes rule* and the term $P(D \mid H, B)$ as the *likelihood*, $L(D)$.

In some cases, the probability of H is not influenced by the knowledge of D, so it is said that H and D are *independent* given some background B, therefore $P(H \mid D, B) = P(H \mid B)$. In the case in which A and B are independent, the product rule can be simplified to: $P(A, B \mid C) = P(A \mid C)P(B \mid C)$, and this can be generalized to N mutually independent propositions:

$$P(A_1, A_2, \ldots, A_N \mid B) = P(A_1 \mid B)P(A_2 \mid B) \cdots P(A_N \mid B) \quad (2.4)$$

Dependence and independence are symmetric relations–if A is dependent on B, then B is dependent on A, and if A is independent of B, then B is independent of A.

If two propositions are independent given only the background information, they are *marginally* independent; however if they are independent given some additional evidence, E, then they are *conditionally* independent: $P(H \mid D, B, E) = P(H \mid B, E)$. For example, consider that A represents the proposition *watering the garden*, B the *weather forecast* and C *raining*. Initially, watering the garden is not independent of the weather forecast; however, once we observe rain, they become independent. That is, $P(A \mid B, C) = P(A \mid C)$.

Probabilistic graphical models are based on these conditions of marginal and conditional independence.

---

[1] It is commonly written $P(H)$ without explicit mention of the conditioning information. In this case we assume that there is still some context under which probabilities are considered even if it is not written explicitly.

The probability of a conjunction of N propositions, that is $P(A_1, A_2, \ldots, A_N \mid B)$, is usually called the *joint* probability. If we generalize the product rule to N propositions we obtain what is known as the *chain* rule:

$$P(A_1, A_2, \ldots, A_N \mid B) = P(A_1 \mid A_2, A_3, \ldots, A_N, B)$$
$$P(A_2 \mid A_3, A_4, \ldots, A_N, B) \cdots P(A_N \mid B) \qquad (2.5)$$

Thus, the joint probability of N propositions can be obtained by this rule. Conditional independence relations between the propositions can be used to simplify this product; that is, for instance if A_1 and A_2 are independent given $A_3, \ldots, A_N, B$, then the first term in Eq. 2.5 can be simplified to $P(A_1 \mid A_3, \ldots, A_N, B)$.

Another important relation is the rule of *total probability*. Consider a partition, $B = \{B_1, B_2, \ldots B_n\}$, on the sample space Ω, such that $\Omega = B_1 \cup B_2 \cup \cdots \cup B_n$ and $B_i \cap B_j = \emptyset$. That is, B is a set of mutually exclusive events that cover the entire sample space. Consider another event A; A is equal to the union of its intersections with each event $A = (B_1 \cap A) \cup (B_2 \cap A) \cup \cdots \cup (B_n \cap A)$. Then, based on the axioms of probability and the definition of conditional probability we can derive the rule of total probability:

$$P(A) = \sum_i P(A \mid B_i) P(B_i) \qquad (2.6)$$

This is very useful, as often we cannot assess $P(A)$ directly, but we can through this decomposition. It is generally easier to assess conditional probabilities such as $P(A \mid B_i)$, which are tied to specific contexts, rather than $P(A)$, which is not attached to a context

Given the total probability rule, we can rewrite Bayes rule as (omitting the background term):

$$P(B \mid A) = \frac{P(B)P(A \mid B)}{\sum_i P(A \mid B_i) P(B_i)} \qquad (2.7)$$

This last expression is commonly known as Bayes Theorem. Bayes' theorem is very important given that in many cases we know or can easily determine $P(B \mid A)$ (the probability that a piece of evidence will occur, given that our hypothesis is correct), but it's much harder to estimate $P(A \mid B)$ (the probability of the hypothesis being correct, given that we obtain a piece of evidence). The latter is the question that we most often want to answer in the real world; generally, we want to update our belief in some hypothesis after some evidence has occurred. For example, we want to estimate the probability of certain disease given some symptom, $P(D \mid S)$, and usually there are statistics that can help us determine the probability of the symptom given the disease, $P(S \mid D)$. Given the prevalence of the disease (prior probability), we can apply Bayes' theorem to obtain the desired probability.

2.2.2   Random Variables

A variable is any property or descriptor that can take multiple values. A variable can be either discrete or continuous. Discrete variables can take one of a finite or countably infinite set of values. Continuous variables can take any one of an infinite set of values on a continuous scale.

If we consider a finite set of exhaustive and mutually exclusive propositions,[2] then a discrete variable X can represent this set of propositions, such that each value x_i of X corresponds to one proposition. If we assign a numerical value to each proposition x_i, then X is a *discrete random variable*. For example, the outcome of the toss of a die is a discrete random variable with 6 possible values 1, 2, ..., 6. The probabilities for all possible values of X, $P(X)$ is the probability distribution of X. Considering the die example, for a fair die the probability distribution will be:

x	1	2	3	4	5	6
$P(x)$	1/6	1/6	1/6	1/6	1/6	1/6

This is an example of a *uniform* probability distribution. There are several probability distributions which have been defined. Another common distribution is the *binomial* distribution. Assume we have an urn with N colored balls, red and black, of which M are red, so the fraction of red balls is $\pi = M/N$. We draw a ball at random, record its color, and return it to the urn, mixing the balls again (so that, in principle, each draw is independent of the previous one). The probability of getting r red balls in n draws is:

$$P(r \mid n, \pi) = \binom{n}{r} \pi^r (1 - \pi)^{n-r}, \tag{2.8}$$

where $\binom{n}{r} = \frac{n!}{r!(n-r)!}$.

This is an example of a binomial distribution which is applied when there are n independent trials, each with two possible outcomes (success or failure), and the probability of success is constant over all trials. There are many other distributions, we refer the interested reader to the additional reading section at the end of the chapter.

There are two important quantities that, in general, help to characterize a probability distribution. The expected value or *expectation* of a discrete random variable is the average of the possible values, weighted according to their probabilities:

$$E(X \mid B) = \sum_{i=1}^{N} P(x_i \mid B)x_i \tag{2.9}$$

---

[2] This means that one and only one of the propositions has a value of TRUE.

The *variance* is defined as the expected value of the square of the variable minus its expectation:

$$\sigma^2(X \mid B) = \sum_{i=1}^{N} P(x_i \mid B)(x_i - E(X))^2 \tag{2.10}$$

Intuitively, the variance gives a measure of how *wide* or *narrow* the probabilities are distributed for a certain random variable. The square root of the variance is known as the standard deviation, which is usually more intuitive as its units are the same as those of the variable.

So far we have considered discrete variables, however, the rules of probability can be extended to continuous variables. If we have a continuous variable X, we can divide it into a set of mutually exclusive and exhaustive intervals, such that $P = (a < X \le b)$ is a proposition, thus, the rules derived so far apply to it. A *continuous random variable* can be defined in terms of a *probability density function*, $f(X \mid B)$, such that:

$$P(a < X \le b \mid B) = \int_{a}^{b} f(X \mid B)dx \tag{2.11}$$

The probability density function must satisfy $\int_{-\infty}^{\infty} f(X \mid B)dx = 1$.

An example of a continuous probability distribution is the *Normal* or Gaussian distribution. This distribution plays an important role in many applications of probability and statistics, as many phenomena in nature have an approximately Normal distribution; it is also prevalent in probabilistic graphical models due to its mathematical properties.

A Normal distribution is denoted as $N(\mu, \sigma^2)$, where μ is the *mean* (center) and σ is the *standard deviation* (spread); and it is defined as:

$$f(X \mid B) = \frac{1}{\sigma\sqrt{2\pi}} e^{\left(-\frac{1}{2\sigma^2}(x-\mu)^2\right)} \tag{2.12}$$

where e is exponential function. The density function of the Gaussian distribution is depicted in Fig. 2.1.

Another important continuous distribution is the Exponential distribution; as an example, the time that it takes for a certain piece of equipment to fail is usually modeled by an exponential distribution. The exponential distribution is denoted as $Exp(\beta)$; it has a single parameter $\beta > 0$, and it is defined as:

$$f(X \mid B) = \frac{1}{\beta} e^{-x/\beta}, x > 0 \tag{2.13}$$

An example of an exponential density function is shown in Fig. 2.2.

It is common to represent probability distributions, particularly for continuous variables, using the cumulative distribution function, F. The cumulative distribution

Fig. 2.1 Probability density function of the Gaussian distribution

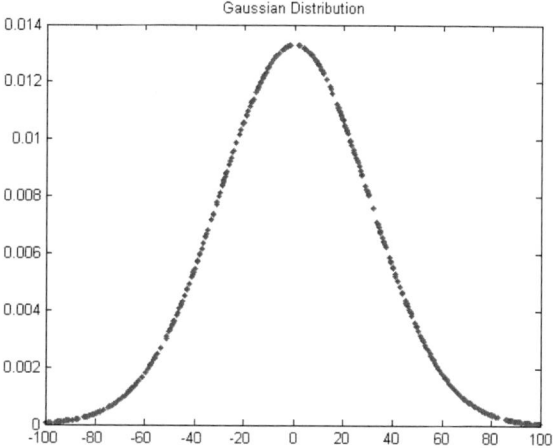

Fig. 2.2 Probability density function of the exponential distribution

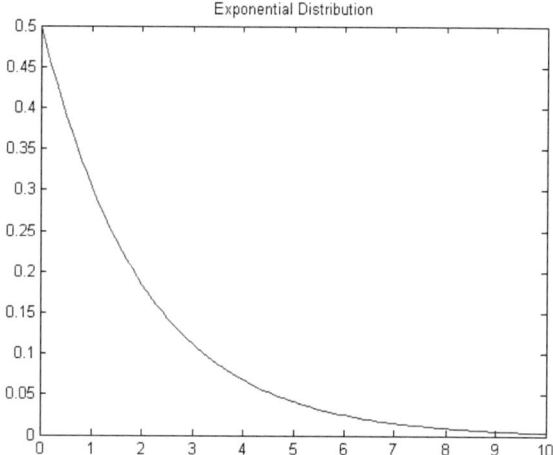

function of a random variable, X, is the probability that $X \leq x$. For a continuous variable, it is defined in terms of the density function as:

$$F(X) = \int_{-\infty}^{x} f(X) \tag{2.14}$$

The following are some properties of cumulative distribution functions:

- In the interval [0, 1]: $0 \leq F(X) \leq 1$
- Non-decreasing: $F(x_1) \leq F(x_2)$ if $x_1 < x_2$
- Limits: $lim_{x \to -\infty} F(X) = 0$ and $lim_{x \to \infty} F(X) = 1$

In the case of discrete variables, the cumulative probability, $P(X \le x)$ is defined as:

$$\mathbf{P}(x) = \sum_{x=-\infty}^{X=x} P(X) \tag{2.15}$$

It has similar properties as the cumulative distribution function.

2.2.2.1 Two Dimensional Random Variables

The concept of a random variable can be extended to two, or more dimensions. Given two random variables, X and Y, their joint probability distribution is defined as $P(x, y) = P(X = x \land Y = y)$. For example, X might represent the number of products completed in one day in product line one, and Y the number of products completed in one day in product line two, thus $P(x, y)$ corresponds to the probability of producing x products in line one and y products in line two. $P(X, Y)$ must follow the axioms of probability, in particular: $0 \le P(X, Y) \le 1$ and $\sum_X \sum_Y P(X, Y) = 1$.

The distribution for two-dimensional discrete random variables (known as the *bivariate* distribution) can be represented in a tabular form. For instance, consider the example of the two product lines, and assume that line one (X) may produce 1, 2 or 3 products per day, and line two (Y), 1 or 2 products. Then, a possible joint distribution, $P(X, Y)$, is shown in Table 2.1.

Given the joint probability distribution, $P(X, Y)$, we can obtain the distribution for each individual random variable, what is known as the marginal probability:

$$P(x) = \sum_y P(X, Y); \ P(y) = \sum_x P(X, Y) \tag{2.16}$$

For instance, if we consider the joint distribution of Table 2.1, we can obtain the marginal probabilities for X and Y. For example, $P(X = 2) = 0.3 + 0.1 = 0.4$ and $P(Y = 1) = 0.1 + 0.3 + 0.3 = 0.7$.

We can also calculate the conditional probabilities of X given Y and vice-versa:

$$P(X \mid Y) = P(X, Y)/P(Y); \ P(Y \mid X) = P(X, Y)/P(X) \tag{2.17}$$

Following the example in Table 2.1:

$$P(X = 3 \mid Y = 1) = P(X = 3, Y = 1)/P(Y = 1) = 0.3/0.7 = 0.4286$$

Table 2.1 An example of a two-dimensional discrete probability distribution.

	X=1	X=2	X=3
Y=1	0.1	0.3	0.3
Y=2	0.2	0.1	0

The concept of independence can be applied to two-dimensional random variables. Two random variables, X, Y are independent if their joint probability distribution is equal to the product of their marginal distributions (for all values of X and Y):

$$P(X, Y) = P(X)P(Y) \rightarrow Independent(X, Y) \tag{2.18}$$

The covariance of X and Y is the expectation of the product $(X - E(X))(Y - E(Y))$:

$$\sigma_{XY} = E[(X - E(X))(Y - E(Y))] \tag{2.19}$$

It measures the degree to which X and Y covary, that is, the degree to which the two variables *linearly* vary together.

Another useful measure is called *correlation*; it is a measure of the degree of linear relation between two random variables, X, Y, and is defined as:

$$\rho(X, Y) = E\{[X - E(X)][Y - E(Y)]\}/(\sigma_X \sigma_Y) \tag{2.20}$$

where $E(X)$ is the expected value of X and σ_X its standard deviation. The correlation is in the interval $[-1, 1]$; a positive correlation indicates that as X increases, Y tends to increase; and a negative correlation that as X increases, Y tends to decrease.

Note that a correlation of zero does not necessarily imply independence, as the correlation only measures a linear relationship. So it could be that X and Y have a zero correlation but are related through a higher order function, and thus they are not independent.

2.2.3 Regression

We often want to predict the value of one variable, Y, based on the value of another variable, X. If we want to make the prediction directly from the data, we apply *regression*; a linear function that takes observed values of X as input and gives values of Y as output, such that the square error between the predicted and actual values of Y is minimized. The least squares regression line is the line for which the sum of the squared vertical distances of the points on the scatter plot from the line is minimized.

In general, we write the regression equation for Y on X as:

$$y = a + bx \tag{2.21}$$

The slope, b, can be obtain based on the covariance:

$$b = R_{YX} = \sigma_{XY}/\sigma_X^2 \tag{2.22}$$

The slope of the regression line can be positive, negative, or zero. If it is positive, X and Y have a positive correlation, meaning that as the value of X gets higher, the value of Y gets higher; if it is negative, X and Y have a negative correlation, meaning that as the value of X gets higher, the value of Y gets lower; if it is zero (a horizontal line), X and Y have no linear correlation. If two variables are correlated, whether positively or negatively, they are dependent.

2.3 Overview of Graph Theory

A *graph* provides a compact way to represent binary relations between a set of objects. For example, consider a set of cities in a certain region, and the roads that connect these cities. Then a map of the region is essentially a graph, in which the object are the cities and the direct roads between pairs of cities are the relations. Graphs are usually represented graphically. Objects are represented as circles or ovals, and relations as lines or arrows; see Fig. 2.3. There are two basic types of graphs: *undirected graphs* and *directed graphs*.

Given V, a non empty set, a binary relation $E \subseteq V \times V$ on V is a set of ordered pairs, (V_j, V_k), such that $V_j, V_k \in V$. A *directed graph* or *digraph* is an ordered pair, $G = (V, E)$, where V is a set of vertices or nodes and E is a set of arcs that represent a binary relation on V; see Fig. 2.3b. Directed graphs represent anti-symmetric relations between objects, for instance the "parent" relation.

An *undirected graph* is an ordered pair, $G = (V, E)$, where V is a set of vertices or nodes and E is a set of edges that represent symmetric binary relations: $(V_j, V_k) \in E \rightarrow (V_k, V_j) \in E$; see Fig. 2.3a. Undirected graphs represent symmetric relations between objects, for example, the "brother" relation.

If there is an edge $E_i(V_j, V_k)$ between nodes j and k, then V_j is adjacent to V_k. In an undirected graph, the *degree* of a node is the number of edges that are incident in that node. In Fig. 2.3a, the upper node has a degree of 2 and the two lower nodes have a degree of 1.

In a directed graph, the number of arcs pointing to a node is its *in degree*; the number of edges pointing away from a node is its *out degree*; and the total number of edges that are incident in the node is its *degree* (in degree + out degree). In Fig. 2.3b, the two upper nodes have an in degree of zero and an out degree of two; while the two lower nodes have an in degree of two and an out degree of zero.

Fig. 2.3 Graphs: a undirected graph, **b** directed graph

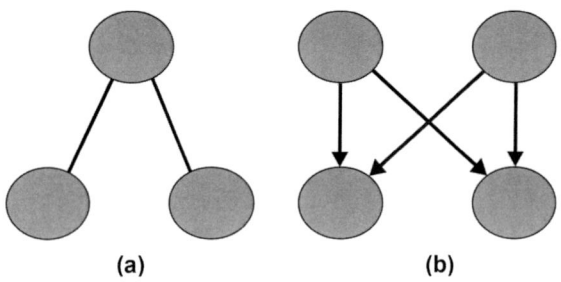

(a)   (b)

Fig. 2.4 Cliques: the 5 cliques in the graph are highlighted

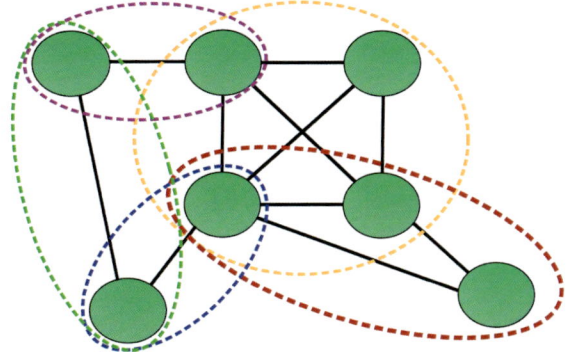

Given a graph $G = (V, E)$, a subgraph $G' = (V', E')$ of G, is a graph such that $V' \subseteq V$ and $E' \subseteq E$, in which each edge in E' is incident on vertices in V'. For example, if we take out the direction of the edges in the graph of Fig. 2.3b (making it an undirected graph), then the graph in Fig. 2.3a is a subgraph of Fig. 2.3b.

A *trajectory* is a sequence of edges, $E_1, E_2, \ldots, E_n$ such that the final vertex of each edge coincides with the initial vertex of the next edge in the sequence (except for the final vertex); that is, $E_i(V_j, V_k)$, $E_{i+1}(V_k, V_l)$, for $i = 1$ to $i = n - 1$. A *circuit* is a trajectory such that the final vertex coincides with the initial one, i.e., it is a "closed trajectory".

An important type of graph is a *Directed Acyclic Graph* (DAG). A DAG is a directed graph that has no directed circuits (a directed circuit is a circuit in which all edges in the sequence follow the directions of the arrows). For instance, Fig. 2.3b is a DAG.

2.3.1   Cliques

A *complete graph* is a graph, G_c, in which each pair of nodes is adjacent; that is, there is an edge between each pair of nodes. A *complete set*, W_c is a subset of G that induces a complete subgraph of G. It is a subset of vertices of G so that each pair of nodes in this subgraph is adjacent. A *clique*, C, is a subset of graph G such that it is a complete set that is maximal; that is, there is no other complete set in G that contains C.

The graph in Fig. 2.4 has 5 cliques, one with four nodes, one with three nodes and three with two nodes. Notice that every node in a graph is part of at least one clique; thus, the set of cliques of a graph, G, always covers G.

2.3.2   Trees

We will discuss two types of trees: undirected and directed. An undirected tree is a connected graph that does not have simple circuits. There are two classes of vertices

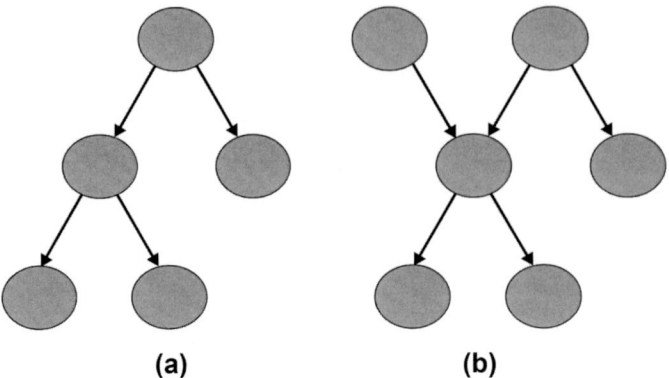

Fig. 2.5 a A rooted tree. **b** A polytree

or nodes in an undirected tree: (i) leaf or terminal nodes, with degree one; (ii) internal nodes, with degree greater than one. Some basic properties of a tree are:

- There is a simple trajectory between each pair of vertices.
- The number of vertices, $| V |$, is equal to the number of edges, $| E |$ plus one: $| V |=| E | +1$.
- A tree with two or more vertices has at least two leaf nodes.

A directed tree is a connected directed graph such that there is only a single directed trajectory between each pair of nodes (it is also known as a singly connected directed graph). There are two types of directed trees: (i) a rooted tree (or simply a tree), (ii) a polytree. A rooted tree has a single node with an in degree of zero (the root node) and the rest have in degree of one. A polytree might have more than one node with in degree zero (roots), and certain nodes (zero or more) with in degree greater than one (called multi-parent nodes). If we take out the direction of the edges in a polytree, it transforms into an undirected tree. We can think of a tree as a special case of a polytree. Examples of a rooted tree and a polytree are shown in Fig. 2.5.

Some terminology for directed trees is the following.

Root: a node with in degree equal to zero.
Leaf: a node with out degree equal to zero.
Internal node: a node with out degree greater than zero (not a leaf node).
Parent/Child: if there is a directed arc from A to B, A is parent of B and B is a child of A.
Brothers: two or more nodes are brothers if they have the same parent.
Ascendants/Descendants: if there is a directed trajectory from A to B, A is an ascendant of B and B is a descendant of A.
Subtree with root A: a subtree with A as its root.
Subtree of A: a subtree with a child of A as its root.

K-ary Tree: a tree in which each internal node has at most K children. It is a regular
    tree if each internal node has K children.
Binary Tree: a tree in which each internal node has at most two children.

2.4 Bayesian Networks

Bayesian networks are directed graphical models that represent the joint distribution
of a set of random variables. An example of a hypothetical medical Bayesian network
is shown in Fig. 2.6. In this graph, the nodes represent random variables and the arcs
direct dependencies between variables. The structure of the graph encodes a set of
conditional independence relations between the variables. For instance, the following
conditional independencies can be inferred from the example:

- *Fever* is independent of *Body ache* given *Flu* (common cause).
- *Fever* is independent of *Unhealthy food* given *Typhoid* (indirect cause).
- *Typhoid* is independent of *Flu* when *Fever* is NOT known (common effect). Knowing Fever makes Typhoid and Flu dependent –for example, if we know that someone has Typhoid and Fever, this *diminishes* the probability of having the Flu.

  In addition to the structure, a Bayesian network considers a set of local parameters,
which are the conditional probabilities for each variable given its parents in the
graph. For example, the conditional probability of Fever given Flu and Typhoid,
$P(Fever \mid Typhoid, Flu)$. Thus, the joint probability of all the variables in the
network can be represented based on these local parameters; this usually implies an
important saving in the number of required parameters.
  Given a Bayesian network (structure and parameters) we can answer several probabilistic queries. For instance, for the previous example: What is the probability of

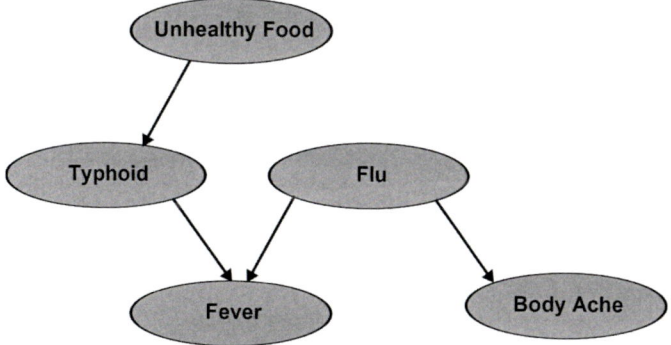

Fig. 2.6 A simple, hypothetical example of a medical Bayesian network

Fig. 2.7 A Bayesian network. The nodes represent random variables and the arcs direct dependencies

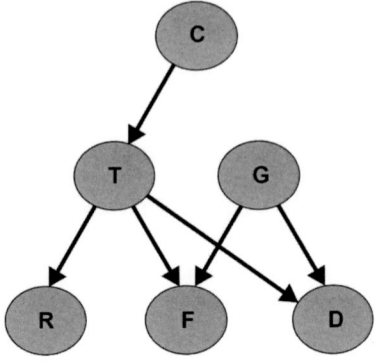

Fever given Flu? Which is more probable, Typhoid or Flu, given Fever and Unhealthy Food?

2.4.1 Representation

A Bayesian network (BN) represents the joint distribution of a set of n (discrete) variables, $X_1, X_2, \ldots, X_n$, as a directed acyclic graph (DAG) and a set of conditional probability tables (CPTs). Each node, that corresponds to a variable, has an associated CPT that contains the probability of each state of the variable given its parents in the graph. The structure of the network implies a set of conditional independence assertions, which give power to this representation.

Figure 2.7 depicts an example of a simple BN. The structure of the graph implies a set of conditional independence assertions for this set of variables. For example, R is conditionally independent of C, G, F, D given T, that is:

$$P(R \mid C, T, G, F, D) = P(R \mid T) \qquad (2.23)$$

Conditional independence assertions can be verified directly from the structure of a BN using a criteria called *D–separation*. Before we define it, we consider the 3 basic BN structures for 3 variables and 2 arcs:

- Sequential: $X \rightarrow Y \rightarrow Z$.
- Divergent: $X \leftarrow Y \rightarrow Z$.
- Convergent: $X \rightarrow Y \leftarrow Z$.

In the first two cases, X and Z are conditionally independent given Y, however in the third case this is not true. This last case, called *explaining away*, corresponds intuitively to having two *causes* with a common *effect*; knowing the effect and one of the causes, alters our belief in the other cause. These cases can be associated with the separating node, Y, in the graph. Thus, depending on the case, Y is sequential, divergent or convergent.

D-Separation

Given a graph G, a set of variables A is conditionally independent of a set B given a set C, if there is no trajectory[3] in G between A and B such that:

1. All convergent nodes are or have descendants in C.
2. All other nodes are outside C.

For instance, for the BN in Fig. 2.7, R is independent of C given T, but T and G are not independent given F.

We will use the following notation to indicate conditional independence: $I(X, Y, Z)$, which indicates that X and Z are independent given Y; where X, Y, Z can be sets of variables. For instance, $I(C, T, R)$ and $I(C, TF, G)$ in Fig. 2.7.

Parameters

To complete the specification of a BN, we need to define its parameters. In the case of a BN, these parameters are the conditional probabilities of each node given its parents in the graph. If we consider discrete variables:

- Root nodes: vector of marginal probabilities.
- Other nodes: conditional probability table (CPT) of the variable given its parents in the graph.

Figure 2.8 shows some of the CPTs of the BN in Fig. 2.7. In case of continuous variables we need to specify a function that relates the density function of each variable to the density of its parents (for example, Kalman filters consider Gaussian distributed variables and linear functions).

2.4.2 Inference

Probabilistic inference consists of *propagating* the effects of certain evidence in a Bayesian network to estimate its effect on the unknown variables. That is, by knowing the values for some subset of variables in the model, the posterior probabilities of the other variables are obtained. The subset of unknown variables could be empty, in this case we obtain the prior probabilities of all the variables.

There are basically two variants of the inference problem in BNs. One is obtaining the posterior probability of a single variable, H, given a subset of known (instantiated) variables, $\mathbf{E}$, that is $P(H \mid \mathbf{E})$. Specifically, we are interested in the marginal

---

[3] In a graph, a *trajectory* is a sequence of edges, $E_1, E_2, \ldots, E_n$ such that the final vertex of each edge coincides with the initial vertex of the next edge in the sequence (except for the final vertex).

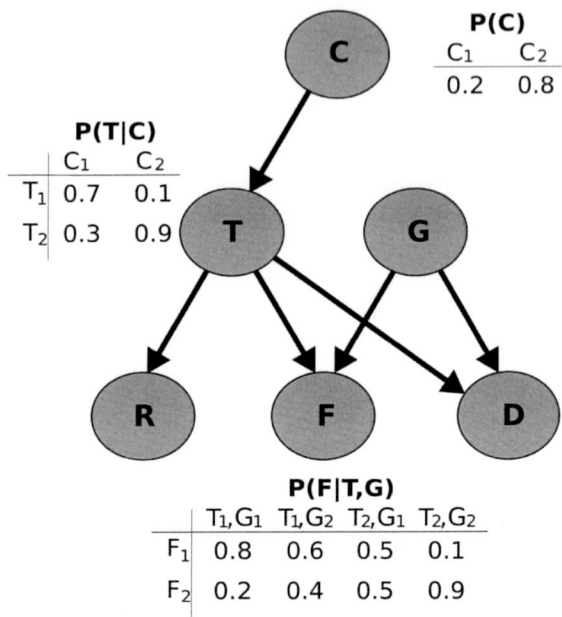

Fig. 2.8 Parameters for the BN in Fig. 2.7. It shows the CPTs for some of the variables in the example: $P(C)$; $P(T \mid C)$; and $P(F \mid T, G)$. We assume in this case that all variables are binary

probabilities of the unknown variables in the model. This is the most common application of BNs, and we will denominate it as *single query inference*.

The second variant consists in calculating the posterior probability of a set of variables, **H** given the evidence, **E**, that is, $P(\mathbf{H} \mid \mathbf{E})$. This is known as *conjunctive query inference*. In principle, it can be solved using single query inference several times by applying the chain rule, making it a more complex problem. For example, $P(A, B \mid \mathbf{E})$ can be written as $P(A \mid \mathbf{E})P(B \mid A, \mathbf{E})$, which requires two single query inferences, and a multiplication. In some applications, it is of interest to know which are the most probable values in the set of hypotheses. That is, $ArgMax_{\mathbf{H}} P(\mathbf{H} \mid \mathbf{E})$. When **H** includes all non observed variables, it is known as the *most probable explanation* (MPE) or the *total abduction* problem. When we are interested in the most likely joint state of some (not all) of the unobserved variables, it corresponds to the *maximum a posteriori probability* (MAP) or *partial abduction* problem.

If we want to solve the inference problem using a direct (brute force) computation (i.e., from the joint distribution), the computational complexity increases exponentially with respect to the number of variables, and the problem becomes intractable even with few variables. Many algorithms have been developed to make this process more efficient, which can be roughly divided into the following classes:

1. Probability propagation (Pearl's algorithm [10]).
2. Variable elimination.
3. Conditioning.
4. Junction Tree.
5. Stochastic simulation.

Fig. 2.9 In a tree–structured
BN, every node (B) divides
the network into two
conditionally independent
subtrees, **E+** and **E−**

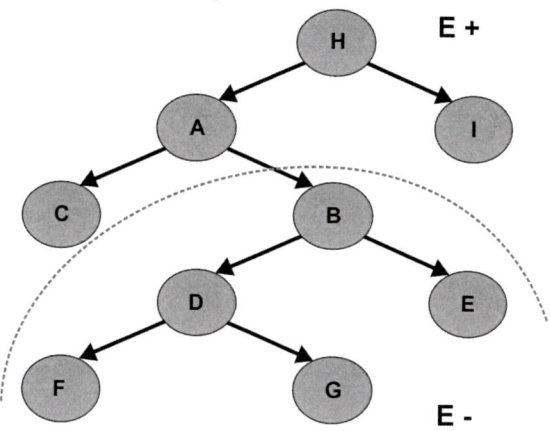

The probability propagation algorithm only applies to singly connected graphs (trees and polytrees[4]), although there is an extension for general networks called *loopy propagation*, this does not guarantee convergence. The other four classes of algorithms work for any network structure, the last one being an approximate technique, while the other three are exact.

In the worst case, the inference problem is *NP-hard* for Bayesian networks [1]. However, there are efficient (polynomial) algorithms for certain types of structures (singly connected networks); while for other structures it depends on the connectivity of the graph. In many applications, the graphs are *sparse*, so in this case there are inference algorithms which are very efficient.

2.4.2.1 Singly Connected Networks: Belief Propagation

We now describe the tree propagation algorithm proposed by Pearl, which provides the basis for several of the most advanced and general techniques.

Given certain evidence, **E** (subset of instantiated variables), the posterior probability for a value i of any variable B, can be obtained by applying the Bayes rule:

$$P(B_i|\mathbf{E}) = P(B_i)P(\mathbf{E}|B_i)/P(\mathbf{E}) \qquad (2.24)$$

Given that the BN has a tree structure, any node divides the network into two independent subtrees. Thus, we can separate the evidence into (see Fig. 2.9):

E–: Evidence of the rooted tree in B.
E+: All other evidence.

---

[4] A polytree is a singly connected DAG in which some nodes have more than one parent; in a directed tree, each node has at most one parent.

Then:

$$P(B_i|\mathbf{E}) = P(B_i)P(\mathbf{E}-, \mathbf{E}+|B_i)/P(\mathbf{E}) \tag{2.25}$$

Given that $\mathbf{E}+$ and $\mathbf{E}-$ are independent, by applying the Bayes rule again, we obtain:

$$P(B_i|\mathbf{E}) = \alpha P(B_i|\mathbf{E}+)P(\mathbf{E}-|B_i) \tag{2.26}$$

where α is a normalization constant. If we define the following terms:

$$\lambda(B_i) = P(\mathbf{E}-|B_i) \tag{2.27}$$

$$\pi(B_i) = P(B_i|\mathbf{E}+) \tag{2.28}$$

Then Eq. 2.26 can be written as:

$$P(B_i|\mathbf{E}) = \alpha\pi(B_i)\lambda(B_i) \tag{2.29}$$

Equation 2.29 is the basis for a distributed propagation algorithm to obtain the posterior probability of all non-instantiated nodes. The computation of the posterior probability of any node B is decomposed into two parts: (i) the evidence coming from the sons of B in the tree (λ), and the evidence coming from the parent of B, (π). We can think of each node B in the tree as a simple processor that stores its vectors $\pi(B)$ and $\lambda(B)$, and its conditional probability table, $P(B \mid A)$. The evidence is propagated via a message passing mechanism, in which each node sends the corresponding messages to its parent and sons in the tree.

Next, we derive the equations for the messages. First, the λ messages that are propagated from the leaves to the root. Given that the sons of B are conditionally independent given B:

$$\lambda(Bi) = P(\mathbf{E}-|Bi) = \prod_k P(\mathbf{E_k}- \mid B_i), \tag{2.30}$$

where $\mathbf{E_k}-$ is the evidence coming from the tree rooted in the S^k son of B. By applying the rule of total probability conditioning on S^k we obtain:

$$P(\mathbf{E_k}- \mid B_i) = \sum_j P(\mathbf{E_k}- \mid B_i, S_j^k)P(S_j^k \mid B_i), \tag{2.31}$$

Given that the evidence coming from the tree rooted in the S^k is conditionally independent of B given S^k:

$$P(\mathbf{E_k}- \mid B_i) = \sum_j P(\mathbf{E_k}- \mid S_j^k)P(S_j^k \mid B_i), \tag{2.32}$$

According to the definition of λ, $P(\mathbf{E_k}- \mid S_j^k) = \lambda(S_j^k)$, substituting in the previous equation:

$$P(\mathbf{E_k}- \mid B_i) = \sum_j P(S_j^k \mid B_i)\lambda(S_j^k). \tag{2.33}$$

Now, the π messages propagated from the root to the leaves. We again apply the rule of total probability conditioning on A, the parent of B:

$$\pi(Bi) = P(B_i|\mathbf{E}+) = \sum_j P(B_i \mid \mathbf{E}+, A_j)P(A_j \mid \mathbf{E}+), \tag{2.34}$$

Given that B is conditionally independent of the evidence coming from the rest of the tree, except the subtree with root B (the ascendants and other descendants of A), given A:

$$\pi(B_i) = \sum_j P(B_i \mid A_j)P(A_j \mid \mathbf{E}+), \tag{2.35}$$

$P(A_j \mid \mathbf{E}+)$ corresponds to the probability of A_j given the evidence coming from all the tree except the subtree rooted on B, it can be written based on Eqs. 2.29 and 2.30 excluding the evidence coming B and its descendants:

$$P(A_j \mid \mathbf{E}+) = \alpha\pi(A_j) \prod_{k \neq b} P(\mathbf{E_k}- \mid A_j) = \alpha\pi(A_j) \prod_{k \neq b} \lambda_k(A_j), \tag{2.36}$$

where b indicates the variable B (one of the children of A). Substituting in Eq. 2.35:

$$\pi(B_i) = \sum_j P(B_i \mid A_j)[\alpha\pi(A_j) \prod_{k \neq b} \lambda_k(A_j)]. \tag{2.37}$$

In summary, the message passing mechanism is the following. Every node B sends a message to its parent A:

$$\lambda_B(Ai) = \sum_j P(B_j \mid A_i)\lambda(B_j) \tag{2.38}$$

Each node can receive several λ messages, which are combined via a term by term multiplication for the λ messages received from each son. Therefore, the λ for a node A with m sons is obtained as:

$$\lambda(Ai) = \prod_{j=1}^{m} \lambda_{Sj}(Ai) \tag{2.39}$$

Every node B sends a message to each son S_l:

$$\pi_l(B_i) = \sum_j P(B_i \mid A_j)[\alpha\pi(A_j) \prod_{k \neq b} \lambda_k(A_j)]. \tag{2.40}$$

where k refers to each one of the sons of B.

Fig. 2.10 A simple BN used
in the belief propagation
example

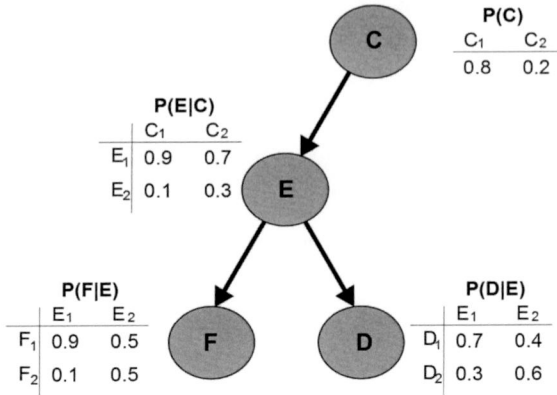

The propagation algorithm starts by assigning the evidence to the known variables, and then propagating it through the message passing mechanism from the leaves until the root of the tree is reached for the λ messages, and from the root until the leaves are reached for the π messages.

For the root and leaf nodes we need to define some initial conditions:

Leaf nodes: If not known, $\lambda = [1, 1, ..., 1]$ (a uniform distribution). If known, $\lambda = [0, 0, ..., 1, ..., 0]$ (one for the assigned value and zero for all other values).
Root node: If not known, $\pi = P(A)$ (prior marginal probability vector). If known, $\pi = [0, 0, ..., 1, ..., 0]$ (one for the assigned value and zero for all other values).

We now illustrate the belief propagation algorithm with a simple example. Consider the BN in Fig. 2.10 with 4 binary variables (each with values *false* and *true*), C, E, F, D, with the CPTs shown in the figure.

Consider that the only evidence is $F = false$. Then the initial conditions for the leaf nodes are: $\lambda_F = [1, 0]$ and $\lambda_D = [1, 1]$ (no evidence). Propagating to the parent node (E) is basically multiplying the λ vectors by the corresponding CPTs:

$$\lambda_F(E) = [1, 0] \begin{bmatrix} 0.9, 0.5 \\ 0.1, 0.5 \end{bmatrix} = [0.9, 0.5]$$

$$\lambda_D(E) = [1, 1] \begin{bmatrix} 0.7, 0.4 \\ 0.3, 0.6 \end{bmatrix} = [1, 1]$$

Then, $\lambda(E)$ is obtained by combining the messages from its two sons:

$$\lambda(E) = [0.9, 0.5] \times [1, 1] = [0.9, 0.5]$$

And now it is propagated to its parent, C:

$$\lambda_E(C) = [0.9, 0.5] \begin{bmatrix} 0.9, 0.7 \\ 0.1, 0.3 \end{bmatrix} = [0.86, 0.78]$$

In this case $\lambda(C) = [0.86, 0.78]$, as C has only one son. In this way, we completed the bottom-up propagation; we will now do it top–down.

Given that C is not instantiated, $\pi(C) = [0.8, 0.2]$, we propagate to its son, E, which also corresponds to multiplying the π vector by the corresponding CPT:

$$\pi(E) = [0.8, 0.2] \begin{bmatrix} 0.9, 0.7 \\ 0.1, 0.3 \end{bmatrix} = [0.86, 0.14]$$

We now propagate to its son D; however, given that E has another son, F, we also need to consider the λ message from this other son, thus:

$$\pi(D) = [0.86, 0.14] \times [0.9, 0.5] \begin{bmatrix} 0.7, 0.4 \\ 0.3, 0.6 \end{bmatrix} = [0.57, 0.27]$$

This finishes the top–down propagation (we do not need to propagate to F as this variable is known). Given the λ and π vectors for each unknown variable, we just multiply them term by term and then normalize to obtain the posterior probabilities:

$$P(C) = [0.8, 0.2] \times [0.86, 0.78] = \alpha[0.69, 0.16] = [0.815, 0.185]$$

$$P(E) = [0.86, 0.14] \times [0.9, 0.5] = \alpha[0.77, 0.07] = [0.917, 0.083]$$

$$P(D) = [0.57, 0.27] \times [1, 1] = \alpha[0.57, 0.27] = [0.68, 0.32]$$

This concludes the belief propagation example.

Probability propagation is a very efficient algorithm for tree structured BNs. The time complexity to obtain the posterior probability of all the variables in the tree is proportional to the *diameter* of the network (the number of arcs in the trajectory from the root to the most distant leaf).

The message passing mechanism can be directly extended to polytrees, as these are also singly connected networks. In this case, a node can have multiple parents, so the λ messages should be sent from a node to all its parents. The time complexity is in the same order as for tree structures.

The propagation algorithm only applies to singly connected networks. Next we will present a general algorithm that applies to any structure.

2.4.2.2   Multiple Connected Networks

There are several classes of algorithms for exact probabilistic inference on multi connected BNs: (i) variable elimination, (ii) conditioning, (iii) junction tree. We will describe the junction tree algorithm that is the most commonly used. (A description of the alternative techniques can be found in the references provided in the additional reading section.)

Junction Tree Algorithm

The junction tree method is based on a transformation of a BN to a junction tree, where each node in this tree is a group or cluster of variables from the original network. Probabilistic inference is performed over this new representation.

The intuition behind the junction tree method is based on a transformation of a BN (which is a directed graph) to a Markov network (undirected graph); and then a clustering of the variables so that the resulting graph is singly connected. Consider a simple BN represented as a chain:

$$A \rightarrow B \rightarrow C \rightarrow D$$

In this case the clusters (cliques) are AB, BC, and CD; and the common variables between the neighbor clusters (separators) are B and C. According to the structure of the BN, the joint probability is $P(A, B, C, D) = P(A)P(B \mid A)P(C \mid B)P(D \mid C)$. Which can be written as:

$$P(A, B, C, D) = P(A)\frac{P(A, B)}{P(A)}\frac{P(B, C)}{P(B)}\frac{P(C, D)}{P(C)} \tag{2.41}$$

Which can be simplified to:

$$P(A, B, C, D) = \frac{P(A, B)P(B, C)P(C, D)}{P(B)P(C)} \tag{2.42}$$

That is basically the product of the probabilities of the clusters divided by the probabilities of the separators; which is the basis of the algorithm.

Next, we describe the junction tree algorithm. The algorithm consists of two phases: (i) transformation of the BN to a junction tree, (ii) probability propagation over the resulting singly connected network.

The transformation proceeds as follows (see Fig. 2.11):

1. Eliminate the directionality of the arcs.
2. Order the nodes in the graph (based on *maximum cardinality* [17]).
3. Moralize the graph (add an arc between pairs of nodes with common children).
4. If necessary, add additional arcs to make the graph *triangulated*.[5]
5. Obtain the *cliques* of the graph (subsets of nodes that are fully connected and are not subsets of other fully connected sets).
6. Build a junction tree in which each node is a clique and its parent is any node that contains all common previous variables according to the ordering.[6]

---

[5] A graph is triangulated if very simple circuit of length greater than three has a *chord* – an edge that connects two vertices in the circuit and is no part of the circuit.
[6] Although a node could have multiple parents, only one is chosen when the junction tree is built.

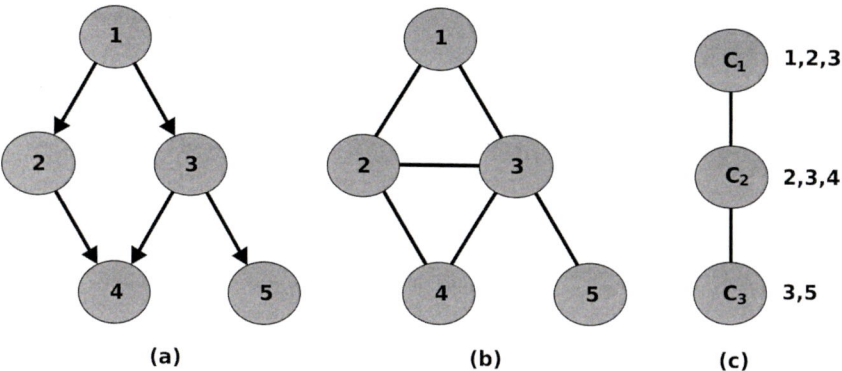

Fig. 2.11 Transformation of a BN to a junction tree: **a** original net, **b** triangulated graph, **c** junction tree

Fig. 2.12 Junction tree of Fig. 2.11 with separator nodes

Ideally, we would like to find a triangulated graph with minimal maximal clique size to make the computations of marginals inside each clique more efficient. However, this is an NP-Hard problem, and certain heuristics for determining the elimination order can be applied.

This transformation procedure guarantees that the resulting junction tree satisfies the *running intersection property*; that is, that common variables with previous cliques are all in one clique. These common variables of neighbor cliques in the junction tree are called *separators*. Given the relevance of these separators, the junction tree is usually drawn including the separator nodes, depicted as rectangles. Figure 2.12 shows the junction tree of Fig. 2.11 including the separator nodes.

Once the junction tree is built, inference is based on probability propagation over the junction tree, analogously as for tree–structured BNs. We must select a root node to begin. Each link between nodes and separators will be used twice during message passing, once in each direction. This is done by propagating messages up from each leaf to the root and then down from the root to the leaves.

The junction tree algorithm can be divided into two stages: preprocessing and propagation. In the preprocessing phase the potentials of each clique are obtained following the next steps:

1. Determine the set of variables for each clique, C_i.
2. Determine the set of variables that are common with the previous (parent) clique, the separators, S_i.
3. Determine the variables that are in C_i but not in S_i: $R_i = C_i - S_i$.
4. Calculate the potential of each clique, clq_i, as the product of the corresponding CPTs: $\psi(clq_i) = \prod_j P(X_j \mid Pa(X_j))$; where X_j are the variables in clq_i.

For example, consider the BN in Fig. 2.11, with cliques: $clq_1 = \{1, 2, 3\}$, $clq_2 = \{2, 3, 4\}$, $clq_3 = \{3, 5\}$. Then the preprocessing phase is:

C: $C_1 = \{1, 2, 3\}$, $C_2 = \{2, 3, 4\}$, $C_3 = \{3, 5\}$.
S: $S_1 = \emptyset$, $S_2 = \{2, 3\}$, $S_3 = \{3\}$.
R: $R_1 = \{1, 2, 3\}$, $R_2 = \{4\}$, $R_3 = \{5\}$.
Potentials: $\psi(clq_1) = P(1)P(2 \mid 1)P(3 \mid 1)$, $\psi(clq_2) = P(4 \mid 3, 2)$, $\psi(clq_3) = P(5 \mid 3)$.

The propagation phase proceeds in a similar way to belief propagation for trees, by propagating λ messages bottom-up and π messages top-down.

Bottom-Up Propagation

1. Calculate the λ message to send to the parent clique: $\lambda(C_i) = \sum_{R_i} \psi(C_i)$.
2. Update the potential of each clique with the λ messages of its sons: $\psi(C_j)' = \lambda(C_i)\psi(C_j)$.
3. Repeat the previous two steps until reaching the root clique.
4. When reaching the root node obtain $P'(C_r) = \psi(C_r)'$.

Top-Down Propagation

1. Calculate the π message to send to each child node i by its parent j: $\pi(C_i) = \sum_{R_j} P'(C_j)$.
2. Update the potential of each clique when receiving the π message of its parent: $P'(C_i) = \psi(C_i)' \frac{\pi(C_i)}{\lambda(C_i)}$.
3. Repeat the previous two steps until reaching the leaf nodes in the junction tree.

At the end of this propagation in both directions, each clique has the joint marginal probability of the variables that conform it. When there is evidence, the potentials for each clique are updated based on the evidence, and the same propagation procedure is followed.

After completing the bidirectional message passing phase, the messages at each clique correspond to its potential functions, which correspond to the joint probability distribution of the nodes in the clique. Because the nodes form a clique, and thus are all connected, to find the probability densities of a subset of these nodes we need to marginalize the resulting potential function over the remaining variables. Thus, the marginal posterior probabilities of each variable are obtained from the clique potentials via marginalization: $P(X) = \sum_{C_i - X} \psi(C_i)$. The method guarantees that these probabilities will be the same not matter from which clique they are calculated. For instance, for the example in Fig. 2.11, we can obtain the probability of variable 2 from C_1 or C_2: $P(2) = \sum_{1,3} \psi(1, 2, 3) = \sum_{3,4} \psi(2, 3, 4)$.

Continuing with the example of the BN in Fig. 2.11, we now illustrate the propagation without evidence.

First C_3 sends a λ message to C_2: $\lambda(C_3) = \sum_5 \psi(3, 5)$. Next we update the potential of C_2: $\psi(C_2)' = \psi(2, 3, 4)\lambda(C_3)$. Then, C_2 sends a λ message to C_1: $\lambda(C_2) = \sum_4 \psi(C_2)'$. And then, we update the potential of C_1: $\psi(C_1)' = \psi(1, 2, 3)\lambda(C_2)$; $P'(C_1) = \psi(C_1)'$. This completes the bottom-up propagation phase.

Now, the top-down propagation starts. C_1 sends a π message to C_2: $\pi(C_2) = \sum_1 P'(C_1)$. Next update the potential of C_2: $P'(C_2) = \psi(C_2)'\frac{\pi(C_2)}{\lambda(C_2)}$. Then, C_2 sends a π message to C_3: $\pi(C_3) = \sum_{2,4} P'(C_2)$. Finally, update the potential of C_3: $P'(C_3) = \psi(C_3)'\frac{\pi(C_3)}{\lambda(C_3)}$.

Let us verify that resulting potentials, $P'(C_i)$, are the joint marginal probability of the corresponding clique. Remember the original potentials of the BN: $\psi(clq_1) = P(1)P(2 \mid 1)P(3 \mid 1)$, $\psi(clq_2) = P(4 \mid 3, 2)$, $\psi(clq_3) = P(5 \mid 3)$.

Bottom-up phase:

$$\lambda(C_3) = \sum_5 P(5 \mid 3)$$

$$\psi(C_2)' = P(4 \mid 2, 3) \sum_5 P(5 \mid 3)$$

$$\lambda(C_2) = \sum_4 P(4 \mid 2, 3) \sum_5 P(5 \mid 3)$$

$$\psi(C_1)' = P(1)P(2 \mid 1)P(3 \mid 1) \sum_4 P(4 \mid 2, 3) \sum_5 P(5 \mid 3)$$

Which can be written as:

$$\psi(C_1)' = \sum_4 \sum_5 P(1)P(2 \mid 1)P(3 \mid 1)P(4 \mid 2, 3)P(5 \mid 3) = \sum_4 \sum_5 P(1, 2, 3, 4, 5)$$

That is the marginalization of 4 and 5 of the joint probability: $P'(C_1) = \psi(C_1)' = P(1, 2, 3)$. Which indeed is the joint of C_1.

Top-down phase:

$$\pi(C_2) = \sum_1 P(1, 2, 3)$$

$$P'(C_2) = \frac{P(4 \mid 2, 3) \sum_5 P(5 \mid 3) \sum_1 P(1, 2, 3)}{\sum_4 P(4 \mid 2, 3) \sum_5 P(5 \mid 3)} = \frac{P(4 \mid 2, 3) \sum_1 P(1, 2, 3)}{\sum_4 P(4 \mid 2, 3)}$$

Given that $\sum_1 P(1, 2, 3) = P(2, 3)$ and $\sum_4 P(4 \mid 2, 3) = 1$, then:

$$P'(C_2) = P(2, 3)P(4 \mid 2, 3) = P(2, 3, 4)$$

Which is the joint of C_2. And finally:

$$\pi(C_3) = \sum_{2,4} P(2, 3, 4)$$

$$P'(C_3) = P(5 \mid 3)\frac{\sum_{2,4} P(2, 3, 4)}{\sum_5 P(5 \mid 3)}$$

Given that $\sum_{2,4} P(2, 3, 4) = P(3)$ and $\sum_5 P(5 \mid 3) = 1$, then:

$$P'(C_3) = P(3)P(5 \mid 3) = P(3, 5)$$

That is the joint probability of C_3. Thus, we have verified that the junction tree algorithm obtains the joint marginals of each cluster or clique, from which we can calculate the single variable's marginals via marginalization. In this case: $P(1) = \sum_{2,3} P'(C_1)$, $P(2) = \sum_{1,3} P'(C_1)$, $P(3) = \sum_{1,2} P'(C_1)$, $P(4) = \sum_{2,3} P'(C_2)$, and $P(5) = \sum_3 P'(C_3)$.

2.4.2.3 Complexity Analysis

In the worst case, probabilistic inference for Bayesian networks is *NP-Hard* [1]. The time and space complexity is determined by what is known as the *tree-width*, and has to do with how close the structure of the network is to a tree. Thus, a tree-structured BN (maximum one parent per variable) has a tree-width of one. A polytree with at most k parents per node has a tree-width of k. In general, the tree-width is determined by how *dense* the topology of the network is, and this affects: (i) the size of the largest factor in the variable elimination algorithm, (ii) the number of variables that need to be instantiated in the conditioning algorithm, and (iii) the size of the largest clique in the junction tree algorithm.

In practice, BNs tend to be sparse graphs, and in this case the exact inference techniques are very efficient even for models with hundreds of variables. In the case of complex networks, an alternative is to use approximate algorithms [8, 17].

2.5 Causal Bayesian Networks

A Causal Bayesian network (CBN) is a directed acyclic graph, G, in which each node represents a variable and the arcs in the graph represent *causal* relations; that is, the relation $A \rightarrow B$ represents some physical mechanism such that the value of B is directly affected by the value of A. If, in a causal graphical model, a variable B

is the child of another variable A, then A is a *direct cause* of B; if B is a descendant of A, then A is a *potential cause* of B.

Causal relations can be interpreted in terms of *interventions*—setting of the value of some variable or variables by an external agent. For example, assume that A stands for a water sprinkler (OFF/ON) and B represents if the grass is wet (FALSE/TRUE). If the grass (B) is originally not WET and the sprinkler is set to ON by an intervention, then B will change to TRUE.

As in BNs, if there is an arc from A to B (A is a direct cause of B), then A is a parent of B, and B is a child of A. Given any variable X in a CBN, $Pa(X)$ is the set of all parents of X. Also, similarly to BNs, when the direct or immediate causes—parents—of a variable are known, the more remote causes (or ancestors) are irrelevant. For example, once we know that the grass is WET, this makes it SLIPPERY, no matter how the grass became wet (we turned on the sprinkler or it rained).

Causal networks represent stronger assumptions than Bayesian networks, as all the relations implied by the network structure should correspond to causal relations. Thus, all the parent nodes, $Pa(X)$, of a certain variable, X, correspond to the direct causes of X. This means that if any of the parent variables of X, or any combination of them, is set to a certain value via an intervention, this will have an effect on X. In a CBN, a variable which is a root node (variable with no parents) is called *exogenous*, and all other variables are *endogenous*.

A simple example of a CBN is depicted in Fig. 2.13, which basically encodes the following causal relations: (i) *Sprinkler* causes *Wet*, (ii) *Rain* causes *Wet*, (iii) *Wet* causes *Slippery*. In this case, *Sprinkler* and *Rain* are exogenous variables, and *Wet* and *Slippery* are endogenous variables.

If a $P(\mathbf{X})$ is the joint probability distribution of the set of variables $\mathbf{X}$, then we define $P_\mathbf{y}(\mathbf{X})$ as the distribution resulting from setting the value for a subset of variables, $\mathbf{Y}$, via an intervention. This can be represented as $do(\mathbf{Y} = \mathbf{y})$, where $\mathbf{y}$ is a set of constants. For instance, given the CBN of Fig. 2.13, if we set the sprinkler to ON, $do(Sprinkler = ON)$, then the resulting distribution will be denoted as $P_{Sprinkler=ON}(Sprinkler, Rain, Wet, Slippery)$.

Fig. 2.13 A simple example of a CBN which represents the relations: *Sprinkler* causes *Wet*, *Rain* causes *Wet*, and *Wet* causes *Slippery*. Example taken from Pearl [11]

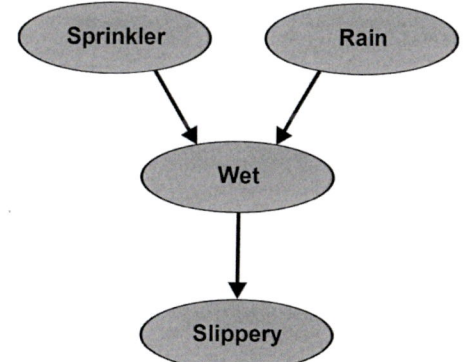

Formally, a Causal Bayesian Network can be defined as follows [11]:

A CBN G is a directed acyclic graph over a set of variables $\mathbf{X}$ that is compatible with all the distributions resulting from interventions on $\mathbf{Y} \subseteq \mathbf{X}$, in which the following conditions are satisfied:

1. The probability distribution $P_y(\mathbf{X})$ resulting from an intervention is Markov compatible with the graph G; that is, it is equivalent to the product of the conditional probability of each variable $X \in G$ given its parents: $P_y(\mathbf{X}) = \prod_{X_i} P(X_i \mid Pa(X_i))$.
2. The probability of all the variables that are part of an intervention is equal to one for the value it is set to: $P_y(X_i) = 1$ if $X_i = x_i$ is consistent with $Y = y$, $\forall X_i \in \mathbf{Y}$.
3. The probability of each of the remaining variables that are not part of the intervention is equal to the probability of the variable given its parents and it is consistent with the intervention: $P_y(X_i \mid Pa(X_i)) = P(X_i \mid Pa(X_i)), \forall X_i \notin \mathbf{Y}$.

Given the previous definition, and in particular the fact that the probability of the variables that are in the intervention set is equal to one (condition 2), the joint probability distribution can be calculated as a *truncated* factorization:

$$P_y(\mathbf{X}) = \prod_{X_i \notin \mathbf{Y}} P(X_i \mid Pa(X_i)) \tag{2.43}$$

such that all X_i are consistent with the intervention $\mathbf{Y}$.

Another consequence of the previous definition, is that once all the parents of a variable X_i are set by an intervention, setting any other variable does not affect the probability of X_i:

$$P_{(Pa(X_i), \mathbf{W})}(X_i) = P_{Pa(X_i)}(X_i) \tag{2.44}$$

such that $\mathbf{W} \cap (X_i, Pa(X_i)) = \emptyset$.

Considering again the example in Fig. 2.13, if we make the grass wet by any mean, $do(Wet = TRUE)$, then the probability of *Slippery* is not affected by *Rain* or *Sprinkler*.

According to the definition of a causal Bayesian network and the intervention (do) operator we can give a formal definition to the manipulationist interpretation of cause and effect. Given a Causal Graphical Model, G, and P_G its corresponding probability distribution; and let X and Y be variables in the model, X causes Y if $P_G(Y \mid do(X = x)) \neq P_G(Y \mid do(X = x')), x \neq x'$.

Causal Bayesian networks allow us to reason about interventions; however, they are not sufficient for counterfactual reasoning. For this we require structural causal models that are described in the next section.

2.6   Structural Causal Models

A structural causal model (SCM), also known as *functional causal model*, consists of two sets of variables **U** and **V**, where **U** are exogenous variables, meaning that they are external to the model, and **V** are endogenous variables; and a set of *structural equations* of the following form:

$$X_i = f_i(Pa(X_i), U_i), i = 1, ..., n \tag{2.45}$$

where $Pa(X_i)$ are the *parents* of X_i, that is the variables that directly determine the values of X_i; and U_i represent the error or disturbances associated to each equation (could be due to other factors not included in the model). In contrast with algebraic equations, structural equations change meaning under algebraic operations; these preserve the solution of the equations but alter their causal meaning.

All endogenous variable in a causal model are a descendant of at least one exogenous variable. Exogenous variables cannot be descendants of any other variables, and in particular, cannot be a descendant of an endogenous variable; they have no ancestors and are represented as root nodes in the causal graphs.

Every SCM is associated with a graphical causal model, similarly to causal Bayesian networks. The main difference is that instead for defining a conditional probability table for each variable given its parents, a function, f, is defined for each variable given its parents in the graph. Graphical models are important as they provide a more intuitive understanding of causality.

Although these equations can take any form, it is common to restrict them to linear equations, and this type of models are known as *linear structural equation models* (LSEMs):

$$X_i = \sum_{k \neq i} \alpha_{ik} X_k + U_i, i = 1, ..., n \tag{2.46}$$

where α_{ik} are constants. This type of models are commonly used in economics and social sciences.

Remember the example of the CBN is depicted in Fig. 2.13, which encodes the following causal relations: (i) *Sprinkler* causes *Wet*, (ii) *Rain* causes *Wet*, (iii) *Wet* causes *Slippery*. If now we use a linear structural equation model it could be represented as the following set of equations:

$$W = \alpha_1 K + \alpha_2 R + U_1 \tag{2.47}$$

$$S = \alpha_3 W + U_2 \tag{2.48}$$

where K denotes the sprinkler, R rain, W wet, and S slippery; U_1 and U_2 are two independent variables that represent the uncertainty (i.e., noise).

An important assumption in functional causal models is that the noise terms are independent variables; otherwise there could be cofactors that affect two or more equations.

A particular class of SEMs are Gaussian linear models (GLMs), which are particularly important from the point of view of causal discovery, Next, we introduce GLMs.

2.6.1 Gaussian Linear Models

Gaussian Linear Models (GLMs), a common type of functional causal model, are those that consider linear relations between variables and Gaussian noise. This type of models are a particular type of *Structural Equation Models* [14], and are easier to learn from observable data.

In a GLM, we have a linear equation that relates a variable to its direct causes (i.e., parents):

$$Y = \beta_1 X_1 + \beta_2 X_2 + \cdots + \beta_q X_q + N_x, \tag{2.49}$$

where $X_1...X_q$ are the direct causes of Y, $\beta_1...\beta_q$ are constants, and N_x is a Gaussian variable with zero mean. Thus:

$$P(Y \mid X_1, \ldots, X_q) = \mathcal{N}(\beta_1 x_1 + \cdots, +\beta_q x_q; \sigma^2). \tag{2.50}$$

The joint distribution of a GLM can be represented as a multivariate Gaussian distribution over a set of n continuous random variables, $\mathbf{X} = \{X_1, \ldots, X_n\}$, by an n-dimensional mean vector, μ, and a symmetric $n \times n$ covariance matrix, $\Sigma = (\sigma_{ij})$. This parameterization of the multivariate Gaussian distribution is called the covariance form, in which the multivariate Gaussian density function is defined as:

$$P(\mathbf{x}) = \frac{1}{(2\pi)^{p/2}|\Sigma|^{1/2}} e^{\left[-\frac{1}{2}(\mathbf{x}-\mu)^T \Sigma^{-1}(\mathbf{x}-\mu)\right]}. \tag{2.51}$$

For multivariate Gaussians, independence is easy to determine directly from the parameters of the distribution. In concrete, if $\mathbf{X} = \{X_1, \ldots, X_n\}$ have a joint normal distribution $\mathcal{N}(\mu; \Sigma)$. Then X_i and X_j are independent if and only if $\sigma_{i,j} = 0$.

For instance, consider a simple model with two variables, X and Y, and the causal model depicted in Fig. 2.14a. X could represent the amount of rain and Y the level of a dam. So, the equations for this model could be:

$$X = N_x$$

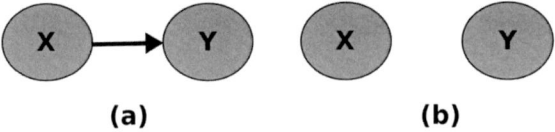

(a)   (b)

Fig. 2.14 **a** A simple example of a GLM which represents the relation: X causes Y. **b** The model resulting after an intervention on Y

$$Y = 8X + N_y$$

with N_x, $N_y \sim N(0, 1)$, which are independent noise variables. This model induces bivariate normal distribution:

$$(X, Y) = \mathcal{N}\begin{pmatrix} 0 \\ 8 \end{pmatrix}\begin{pmatrix} 1 & 8 \\ 8 & 65 \end{pmatrix}$$

Now, let's see what happens if we make interventions in this model. First, consider that we make $X = 10$ (say we make it rain $10\,mm$), so the equations now change to:

$$X = 10$$

$$Y = 80 + N_y$$

And the causal model remains the same as in Fig. 2.14a.

Now, we intervene Y and make it equal to $N(2, 2)$ (assume we can fill the dam via other means). This alters the graphical model, as Y is not influenced by X, so we modified the model as shown in Fig. 2.14b. The joint distribution becomes:

$$(X, Y) = \mathcal{N}\begin{pmatrix} 0 \\ 2 \end{pmatrix}\begin{pmatrix} 1 & 0 \\ 0 & 2 \end{pmatrix}$$

So X and Y are independent. Thus, if we make an intervention in Y this does not affect the value of X; however, as in the previous case, if we intervene X it has a causal effect on Y.

From this example, we can see that an intervention can change the joint distribution of the model; something that does not happen when we measure an observation. Thus, the effect of an intervention in a variable is different from conditioning based on observing a variable. In this example:

$$P_{do(Y=y)}(X) \neq P(X \mid Y = y)$$

2.7 Additional Reading

An excellent book on probability theory from a logical perspective is [4]. Wasserman [18] gives a concise course on probability and statistics oriented for computer science and engineering students. Another accessible introduction to probability and statistics for computer science is [3].

An introduction to Bayesian networks is given in the classic book by Pearl [10]. Sucar [17] provides a general overview of graphical models including Bayesian networks, with the required fundamentals on probability and graph theory, as well as a description of alternative inference algorithms and approximate techniques. Other general books on BNs are [6,9]. A more recent account with emphasis on modeling and inference is given by [2]; it includes a complexity analysis for the different inference techniques. The junction tree algorithm was initially introduced by [7], and the two main architectures are described in [5,15].

Graphical causal modeling was originally introduced in genetics [19]. An accessible introduction to causal models is presented in [13]. Two comprehensive books on graphical causal models are [11,16]. Pearl et al. [12] provides a general introduction to structural equation models and causal inference from an statistical perspective. Pearl et al. [14] gives a general introduction to causality, with emphasis on its relation with statistics and machine learning.

2.8 Exercises

1. Assume that a person has only one of two possible diseases, hepatitis (H) or typhoid (T). There are two symptoms associated to those diseases: headache (D) and fever (F), which could be TRUE or FALSE. Given the following probabilities: $P(T) = 0.5$, $P(D \mid T) = 0.7$, $P(D \mid \neg T) = 0.4$, $P(F \mid T) = 0.9$, $P(F \mid \neg T) = 0.5$. Describe the sampling space and complete the partial probability tables.

2. Given the data for the previous problem, and assuming that the symptoms are independent given the disease, obtain the probability that the person has hepatitis given that she does not have a headache and does have a fever.

3. How will you represent graphically that the two symptoms (headache and fever) are independent given the disease?

4. Given the two dimensional probability distribution in the table below, obtain: (a) $P(X_1)$, (b) $P(Y_2)$, and (c) $P(X_1 \mid Y_1)$.

	Y_1	Y_2	Y_3
X_1	0.1	0.2	0.1
X_2	0.3	0.1	0.2

5. In the previous problem, are X and Y independent?

6. Modify the probabilities of the joint distribution in Problem 4 so that X and Y are independent.

7. Complete the CPTs for the BN in Fig. 2.8 assuming all the variables are binary.

8. Consider the belief propagation example in Sect. 2.4.2.1, obtain the posterior probabilities of all the variables via belief propagation considering that the only evidence is $C = true$.

9. Given the BN structure of Fig. 2.11a and the following CPTs, considering all are binary variables:
$P(1) = [0.6, 0.4]$, $P(2 \mid 1) = \frac{0.3, 0.2}{0.7, 0.8}$, $P(3 \mid 1) = \frac{0.6, 0.1}{0.4, 0.9}$, $P(4 \mid 2.3) = \frac{0.5, 0.1, 0.3, 0.2}{0.5, 0.9, 0.7, 0.8}$, $P(5 \mid 3) = \frac{0.7, 0.7}{0.3, 0.3}$.

(a) Transform it to a junction tree, specify the structure and cliques. (b) Obtain the potentials for each junction. (c) Using the inference algorithm, obtain the marginal probability of each variable.

10. For the previous problem, recalculate the marginal probabilities given the evidence that E has the value false (first value in the CPT).
11. What are the differences between a directed graphical model, such as a Bayesian network, and a causal model?
12. Consider the CBN in Fig. 2.13. Obtain some other alternative Bayesian network models for these 4 variables that represent the same conditional independence relations but which are not necessarily *causal*. How do these networks compare in terms of simplicity and clarity to the original model?
13. Consider the CBN in Fig. 2.13, and that all the variables are binary. Based on your intuition, specify the parameters for the CBN.
14. Repeat the previous problem considering now that the model is represented as a SEM.
15. Assume we have the following knowledge: the amount of exercise and the diet can influence the probability that a person has overweight, and overweight can provoke a stroke. (a) Represent this knowledge as a CBN. (b) Represent it as a SEM.

Acknowledgements Sections 2.2–2.4 are based on [17], used with permission from Springer. Basic principles about causal models are inspired on [11, 12].

References

1. Cooper GF (1990) The computational complexity of probabilistic inference using bayesian networks. Artif. Intell. 42:393–405
2. Darwiche A (2009) Modeling and reasoning with bayesian networks. Cambridge University Press, New York
3. Forsyth D (2018) Probability and statistics for computer science. Springer, Switzerland
4. Jaynes ET (2003) Probability theory: the logic of science. Cambridge University Press, Cambridge, UK
5. Jensen FV, Andersen SK (1990) Approximations in bayesian belief universes for knowledge based systems. In: Proceedings of the sixth conference on uncertainty in artificial intelligence UAI-90, pp 162–169. Elsevier, New York (1990)
6. Jensen FV (2001) Bayesian networks and decision graphs. Springer-Verlag, New York
7. Lauritzen S, Spiegelhalter DJ (1988) Local computations with probabilities on graphical structures and their application to expert systems. J R Stat Soc Ser B 50(2):157–224
8. Murphy KP, Weiss Y, Jordan M (1999) Loopy belief propagation for approximate inference: an empirical study. In: Proceedings of the fifteenth conference on uncertainty in artificial intelligence, pp 467–475. Morgan Kaufmann Publishers Inc.
9. Neapolitan RE (1990) Probabilistic reasoning in expert systems. Wiley, New York
10. Pearl J (1988) Probabilistic reasoning in intelligent systems: networks of plausible inference. Morgan Kaufmann, San Francisco

11. Pearl J (2009) Causality: models, reasoning and inference. Cambridge University Press, New York
12. Pearl J, Glymour M, Jewell NP (2016) Casual inference in statistics: a primer. Wiley, United Kingdom
13. Pearl J, Mackienze D (2018) The book of why. Basic Books
14. Peters J, Janzing D, Scholkpf B (2017) Elements of causal inference. MIT Press
15. Shenoy P, Shafer G (1990) Axioms for probability and belief-function propagation. In: uncertainty in artificial intelligence, vol 4, pp 169–198. Elsevier, New York (1990)
16. Spirtes P, Glymour C, Scheines R (2000) Causation, prediction, and search. MIT Press
17. Sucar LE (2021) Probabilistic Graphical models: principles and applications. Springer Nature, Switzerland
18. Wasserman L (2004) All of statistcs: a concise course in statistical inference. Springer-Verlag, New York
19. Wright S (1921) Correlation and causation. J Agric Res 20:557–585

Causal Reasoning

3

Abstract

This chapter presents a general introduction to causal reasoning based on graph-ical causal models. First we introduce causal predictions using causal Bayesian networks, including the front door and back door criteria, and the DO calculus, for dealing with cofactors. Then we describe how to solve counterfactuals based on structural causal models.

3.1 Causal Reasoning

The ultimate aim of many studies is to predict the effects of interventions, i.e.: determine if a vaccine will be effective against certain disease, find out the potential reduction in the crime rate given changes in the judicial system, search for strategies in order to decrease wild fires, etc. As we have seen before, the gold standard for predicting the effect of interventions are randomized control experiments. However, many questions do not lend themselves to randomized controlled experiments; these could be too expensive, unethical or simply impossible (for example, we can not control the weather). An alternative is to build a causal model and derive the results of predicting the effect of interventions from this model, doing *causal reasoning*.

Causal reasoning has to do with answering causal queries from a causal model, known as *causal effect estimation*, and in our particular case, from graphical causal models. Given a causal graphical model, and the causal reasoning techniques we will discuss below, we can obtain causal information from purely observational data, assuming that the causal graph constitutes a valid representation of reality.

There are several types of causal queries we might consider, we will start by analyzing causal predictions, and then we will analyze counterfactuals. Causal predictions can be answered given a causal Bayesian network, but for counterfactuals these are not sufficient, so we use structural causal models for reasoning about counterfactuals.

It is important to recall that the result of an intervention, $P(y \mid do(x))$, is different form conditioning, $P(y \mid x)$; intervening on a variable results in a totally different pattern of dependencies than conditioning on a variable. We will also use a shorter notation for an intervention, $P(y_x)$, which is particularly useful for counterfactuals.

3.2 Prediction

Causal prediction consists of estimating the causal effect of an intervention on a variable or set of variables (causes), on other variables (effects). In general, if we want to estimate the causal effect of some variable X on other variable Y, the result can be affected if there are hidden *cofactors*, that is, common ancestors of X and Y that are not observed; so this should be taken into account in the inference process. Initially we present a simplified version of the prediction procedure without considering cofactors, and later, in Sect. 3.3, we describe the criteria to consider these cofactors via what is called an *adjustment set*.

Consider a causal Bayesian network which includes a set of variables: $\mathbf{X_G} = \{\mathbf{X_C}, \mathbf{X_E}, \mathbf{X_O}\}$; where $\mathbf{X_C}$ is a subset of *causes* and $\mathbf{X_E}$ is a subset of *effects*; $\mathbf{X_O}$ are the remaining variables. We want to perform a causal query on the model: What will the effect be on $\mathbf{X_E}$ when setting $\mathbf{X_C} = \mathbf{x_C}$? That is, we want to obtain the probability distribution of $\mathbf{X_E}$ that results from the intervention $\mathbf{X_C} = \mathbf{x_C}$:

$$P_C(\mathbf{X_E} \mid do(\mathbf{X_C} = \mathbf{x_C})) \tag{3.1}$$

To perform causal prediction with a CBN, G, the following procedure is followed:

1. Eliminate all incoming arrows in the graph to all nodes in $\mathbf{X_C}$, thus obtaining a modified CBN, G_r.
2. Fix the values of all variables in $\mathbf{X_C}$, $\mathbf{X_C} = \mathbf{x_C}$.
3. Calculate the resulting distribution in the modified model G_r (via probability propagation as in Bayesian networks).

For example, consider the hypothetical CBN depicted in Fig. 3.1a that represents certain causal knowledge about a stroke. If we want to measure the effect of an

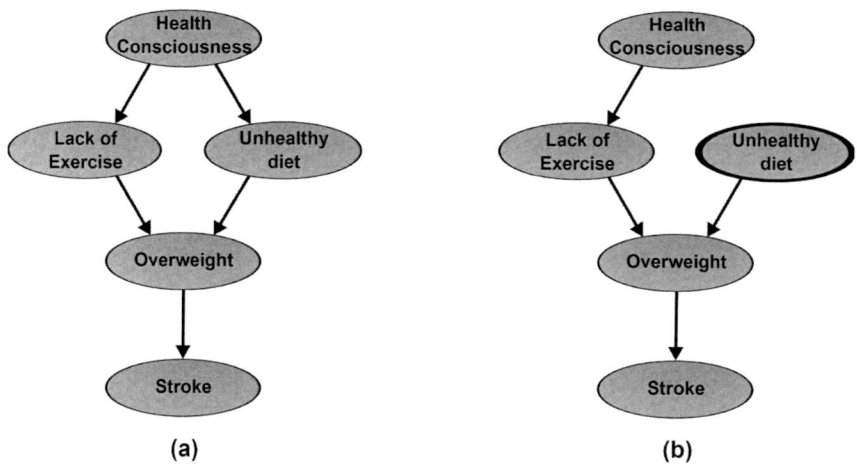

Fig. 3.1 An example of causal prediction. **a** A simple, hypothetical CBN that represents some causal relations related to a stroke. **b** The resulting graphical model obtained by the intervention $do(unhealthy - diet = TRUE)$

"unhealthy diet" in the variable "stroke", then, according to the previous procedure, we eliminate the link from "health consciousness" to "unhealthy diet", resulting in the model in Fig. 3.1b. Then, we should set the value of "unhealthy diet" to $TRUE$, and by probability propagation obtain the distribution of "stroke". If the distribution of "stroke" changes depending on the value of "unhealthy diet", we can conclude according to this model that it does have an effect.

An interesting question is: When is the distribution resulting from an intervention is equal to the distribution resulting from an observation? In mathematical terms, is $P(X_E \mid X_C) = P_C(X_E \mid do(X_C = x_C))$? Both are equal if X_C includes all the parents of X_E and none of its descendants; given that any variable in a BN is independent of its non-descendants given its parents. In other cases they are not necessarily equal, they will depend on other conditions. Thus, in general the joint probability distribution of a CBN is different from the original model after an intervention.

To illustrate the difference between observation and intervention, we will use the following example due to Darwiche [2]. Consider that a new drug to treat cancer is being developed, and data regarding the impact of the drug on survival depending on the stage of the disease has been obtained. Figure 3.2 depicts a causal graph relating the three variables and the associated prior probabilities.

We first consider the case in which we *observe* that someone takes the drug, resulting in the model shown in Fig. 3.3a. In this case the drug seems to be harmful, as the probability of survival is lower, compared to the prior probability (Fig. 3.2). Also the probability of the stage has changed, it is more probable that the person is in an advanced stage.

If we make an intervention on the variable *Drug* we have the modified model in Fig. 3.3b, in which the arc pointing to *Drug* has been deleted. The CPT for the *Drug* variable is modified, replacing the previous CPT conditioned on the *Stage*,

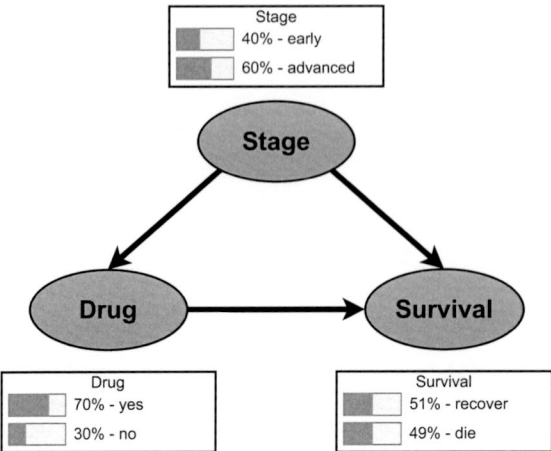

Fig. 3.2 A causal Bayesian network that relates a drug with the probability of survival and the stage of the disease. The prior probabilities for the three variables are shown. Figure taken from [2]

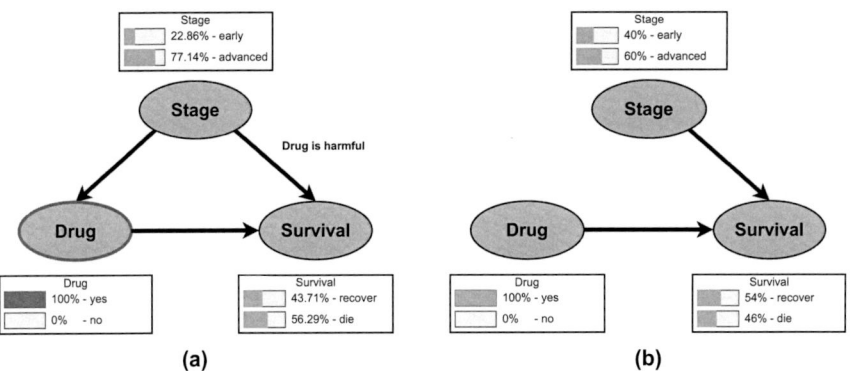

Fig. 3.3 Contrasting the difference between an observation (**a**), and an intervention (**b**), for the example in Fig. 3.2. Figure taken from [2]

with marginal probabilities assigning a probability of 100% to yes, representing an intervention. In this case the effect of the drug is positive, the probability of survival has increased. This example clearly shows the difference between observing and making an intervention; an intervention modifies the joint probability distribution of the graphical model, and in general the results are different.

The previous method of deleting the arcs pointing to the variables which are subject to an intervention is known as the *surgery method*. There are other alternative methods to obtain the same results. The *auxiliary do-node* procedure consists on adding a virtual or auxiliary node pointing to each variable that is intervened. For example, considere the CBN for the stroke example, and an intervention on *Unhealthy Diet*. Figure 3.4 depict the causal graph including the additional *Do* node.

Fig. 3.4 An example of the *auxiliary do-node* procedure. The figure shows the causal Bayesian network for the stroke example incorporating a *Do* node for an intervention on *Unhealthy Diet*

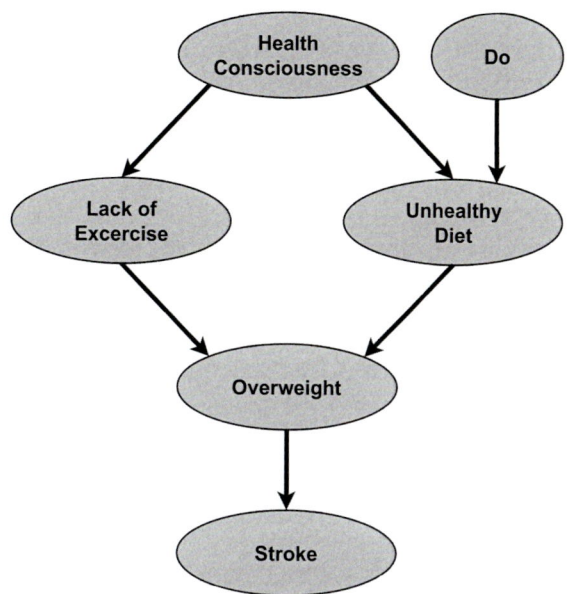

The prior probabilities of the auxiliary variable can be set to any value, for instance a uniform distribution. The CPT for the intervened variable given the *Do* variable must be: (i) equal to the CPT of the original model when there is no intervention, and (ii) will set the intervened variable to a probability of one when making an intervention.

For the example, for no intervention, $Do = NO$ (UD = Unhealthy Diet, HC = Health Consciousness):

$$P(UD \mid HC, Do = NO) = P(UD \mid HC) \tag{3.2}$$

For an intervention, $Do = YES$:

$$P(UD = True \mid HC, Do = YES) = 1; \; P(UD = False \mid HC, Do = YES) = 0 \tag{3.3}$$

By specifying the probabilities in this way, the results will be the same as the previous method when an intervention is performed.

Another way of predicting the effect of an intervention is an algebraic approach based on the surgery method. For this we consider the chain rule to express the joint probability of a BN, but in this case we apply it to the modified model (also known as the truncated graph). Then we eliminate the variables that have been subject to an intervention, and set the values according to the intervention when these appear in the conditional probabilities. Again, considering the same stroke example and an intervention on Unhealthy diet:

$$P_{UD=False}(HC, LE, O, S) = P(HC)P(LE \mid HC)P(O \mid LE, UD = False)P(S \mid O) \tag{3.4}$$

where UD = Unhealthy Diet, HC = Health Consciousness, LE = Lack of Exercise, O = Overweight, and S = Stroke.

3.3 Identification of Causal Effects

An important question is when, given some intervention, a causal effect is *identifiable*. If we have the graphical causal model and all the associated parameters, we can always identify the causal effects. This can be done based on associational probabilities applying the *causal effect rule*:

$$P(y \mid do(x)) = \sum_{\mathbf{z}} P(y \mid x, \mathbf{z}) P(\mathbf{z}) \tag{3.5}$$

where $\mathbf{Z}$ are the parents of X in the causal graph. For this we need to know which are the parents of each variable; that is, to have the appropriate causal structure.

In practice, we might have the causal structure and some data, and want to verify if a causal effect is identifiable. There are several crieria to test for identifiability. Given a CBN, we want to determine the results of an intervention, $P(Y \mid do(X = x))$, in the presence of some cofactors (a cofactor is a variable that has causal links towards the cause and effect variables, in this case X and Y). For example, Z in $X \leftarrow Z \rightarrow Y$. Then, the question is if we can always identify the causal effect from the joint distribution of the variables in the CBN? The answer is "yes" when the covariates $\mathbf{Z}$ contain all the other relevant variables. If some variables are not observed, then the issue of which causal effects are observationally identifiable is considerably trickier.

When there are unobserved (hidden) variables there are cases in which it is impossible to estimate the causal effects given observational probabilities. To illustrate this, consider the causal graph depicted in Fig. 3.5, where X, Y, Z are observed variables and U is hidden. In this case, $P(y \mid do(x))$ and $P(y \mid do(z))$ are identifiable, but $P(x \mid do(z))$ is not identifiable. Two cases in which the causal effect are always identifiable are: (i) when all the parents of the cause are observed, (ii) when all the hidden variables have only one children. In the cases where these two conditions are not satisfied, the effect could be or not identifiable, and for this we can apply the criteria we will cover next.

To determine if a causal effect is identifiable, the basic principle is that we would like to condition (intervene) on the adequate control variables, which will *block*[1] paths linking X and Y other than those which would exist in the altered graph where

---

[1] A path is blocked in a causal graph when a node (variable) or set of nodes are instantiated so the effect of the evidence can not go through this node(s). For this, it is required that at least one collider (node with convergent arcs) is not instantiated, or at least one non-collider is instantiated. For instance, in the causal graph $X \rightarrow Y \rightarrow Z$, if Y is instantiated it will block the path from X to Z; but in the causal graph $X \rightarrow Y \leftarrow Z$, if Y is instantiated, the path is not blocked.

Fig. 3.5 An example of identification of causal effects. In this causal graph, X, Y, Z are observed variables and U is unobserved. $P(y \mid do(x))$ and $P(y \mid do(z))$ are identifiable, but $P(x \mid do(z))$ is not

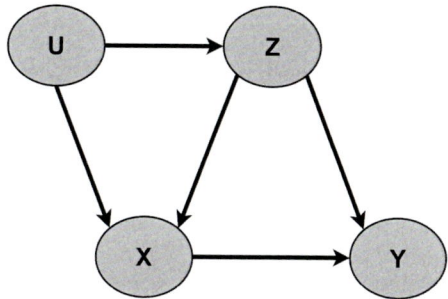

all paths into X have been removed. If other unblocked paths exist, then there could be some confounding of the causal effect of X on Y. There are two basic criteria we can use to get an adequate control; they are called the *back-door criterion* and the *front-door criterion* [4].

The *back-door* and *front-door* criteria are not complete. This is, if they find a solution then the causal effect is identifiable; but if not, it does not mean that it is not identifiable, maybe it could be determine by other means. A more general, and complete approach is the *Do Calculus*. Next we describe the three methdos.

3.3.1   Back Door Criterion

If we desire to know the casual effect of X on Y and have a set of variables $\mathbf{Z}$ as the control, then $\mathbf{Z}$ satisfies the back-door criterion if:

1. $\mathbf{Z}$ blocks every path from X to Y that has an arrow into X (blocks the back door), and
2. No node in $\mathbf{Z}$ is a descendant of X.

Then:

$$P(Y = y \mid do(X = x)) = \sum_z P(Y = y \mid X = x, \mathbf{Z} = \mathbf{z}) P(\mathbf{Z} = \mathbf{z}) \qquad (3.6)$$

where $P(Y = y \mid X = x, \mathbf{Z} = \mathbf{z})$ and $P(\mathbf{Z} = \mathbf{z})$ are observational conditional probabilities. Some examples of the back-door criteria are illustrated in Fig. 3.6. In this case, the causal effect $X \rightarrow Y$ is identifiable usng the back door criteria if we observe $\{Z_1, Z_3\}$ or $\{Z_2, Z_3\}$; but if we do not observe $\{Z_1, Z_2, Z_3\}$ we can not apply the back door criteria. However, it is identifiable by the front door criteria (we will see a similar case later); so it shows that the back door criteria is incomplete.

Fig. 3.6 Given the CBN in the figure, the sets $\{Z_1, Z_3\}$ and $\{Z_2, Z_3\}$ satisfy the back-door criteria for the causal effect $X \rightarrow Y$; but the set $\{Z_3\}$ does not satisfy it as it does not block the trajectory (X, Z_1, Z_3, Z_2, Y)

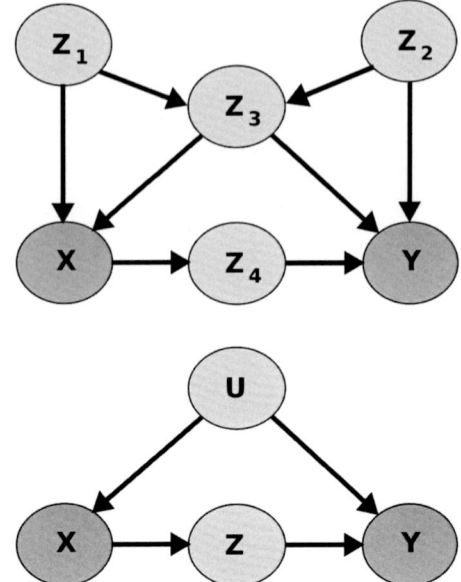

Fig. 3.7 Given the CBN in the figure, the set $\{\mathbf{Z}\}$ satisfies the front-door criteria for the causal effect $X \rightarrow Y$. U is an unobserved variable that is a cofactor of X and Y

3.3.2 Front Door Criterion

If we want to know the causal effect of X on Y and have a set of variables $\mathbf{Z}$ as the control, then $\mathbf{Z}$ satisfies the front-door criterion if:

1. $\mathbf{Z}$ blocks all directed paths from X to Y,
2. there are no unblocked back-door paths from X to $\mathbf{Z}$, and
3. X blocks all back-door paths from $\mathbf{Z}$ to Y.

Then:

$$P(Y = y \mid do(X = x)) = \sum_{\mathbf{z}} P(\mathbf{Z} = \mathbf{z} \mid X = x)$$

$$\sum_{x'} P(Y = y \mid X = x', \mathbf{Z} = \mathbf{z}) P(X = x') \qquad (3.7)$$

To understand the front-door criteria we can analyze it by parts. By clause (1), $\mathbf{Z}$ blocks all directed paths from X to Y, so any causal dependence of Y on X must be mediated by a dependence of Y on $\mathbf{Z}$. Clause (2) says that we can estimate the effect of X on $\mathbf{Z}$ directly; and clause (3) says that X satisfies the back-door criterion for estimating the effect of $\mathbf{Z}$ on Y (the inner sum in Eq. 3.7 is the back-door criterion estimate). An example of the front-door criteria is illustrated in Fig. 3.7.

Both the back-door and front-door criteria are not *complete*; that is, there could be cases in which the causal effects are identifiable and can not be solved by these two

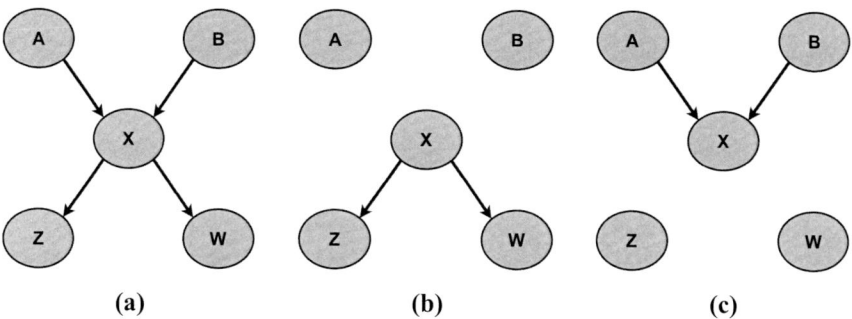

Fig. 3.8 An example of alterations to a graph. **a** Original graph, G. **b** $G_{\overline{X}}$. **c** $G_{\underline{X}}$

approaches. The *DO-calculus* provides a complete procedure for identifying causal effects [4].

3.3.3 Do-Calculus

The Do-Calculus consists of a set of inference rules that allows to transform a probabilistic query about interventions, to other query based on observed quantities (ordinary conditional probabilities). That is, transform a causal query to an expression that can be solved based on statistical information. These rules have been shown to be *complete*; that is, sufficient for deriving all identifiable causal effects [8].

Before we define these rules, we requiere some notation. X, Y and Z are disjoint subsets of nodes of a causal DAG, G. $G_{\overline{X}}$ is the graph derived from G by deleting all the arcs that point to X in G. $G_{\underline{X}}$ is the graph derived from G by deleting all the arcs that emerge from X in G. Figure 3.8 shows examples of these alterations to a graph. $P(y \mid do(x))$ represents the probability of y given an intervention on x, and $I(X, Z, Y)$ corresponds to X and Y being independent given Z (D-separated in the graph).

Given a causal model represented by a DAG, G, and X, Y, Z and W disjoint subsets of variables in G, the rules of the Do Calculus are the following [4]:

Rule 1: Decide if we can ignore an observation (Insertion/deletion of observations)

$$P(y \mid do(x), z, w) = P(y \mid do(x), w) \tag{3.8}$$

if $I(Y, XW, Z)$ in $G_{\underline{X}}$

Rule 2: Decide if we can treat an intervention as an observation (Action/observation exchange)

$$P(y \mid do(x), do(z), w) = P(y \mid do(x), z, w) \tag{3.9}$$

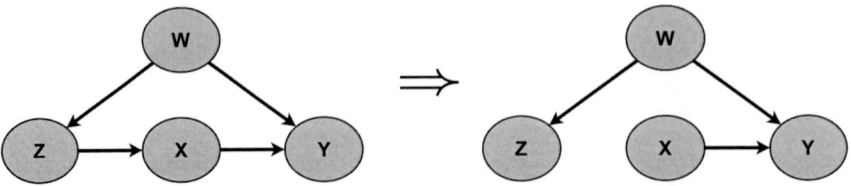

Fig. 3.9 An example of Rule 1

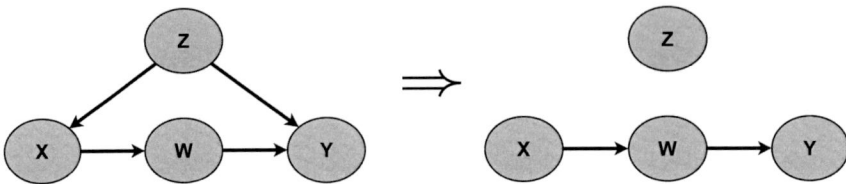

Fig. 3.10 An example of Rule 2

if $I(Y, XW, Z)$ in $G_{\overline{X}\underline{Z}}$

Rule 3: Decide if we can ignore an intervention (Insertion/deletion of actions)

$$P(y \mid do(x), do(z), w) = P(y \mid do(x), w) \qquad (3.10)$$

if $I(Y, XW, Z)$ in $G_{\overline{XZ(W)}}$
where $Z(W)$ is the set of nodes Z that are not ancestors of any W node in $G_{\overline{X}}$.

Let su analyze these rules in more detail. Graphical examples of the application of each rule are depicted in Figs. 3.9, 3.10, and 3.11.

Rule 1 (Fig. 3.9) basically allow us to remove variable Z from the calculation. This is because Z is D-separated (independent) of Y given X, W; it is a generalization of D-separation for interventional distributions. If we consider X to be the empty set, we have:

$$P(y \mid z, w) = P(y \mid w), \qquad (3.11)$$

if $I(Y, W, Z)$ in G, where it is more obvious that this corresponds to D-separation.

Rule 2 (Fig. 3.10) is basically a generalization of the back door criteria. According to Rule 2, interventions can be treated as observations when the causal effect of a variable on the outcome only influences the outcome through directed paths. This is clear if again we consider X to be the empty set, we have:

$$P(y \mid do(z), w) = P(y \mid z, w), \qquad (3.12)$$

if $I(Y, W, Z)$ in $G_{\underline{Z}}$. So we can treat Z as an observation given that there is no direct effect on Y in the modified graph.

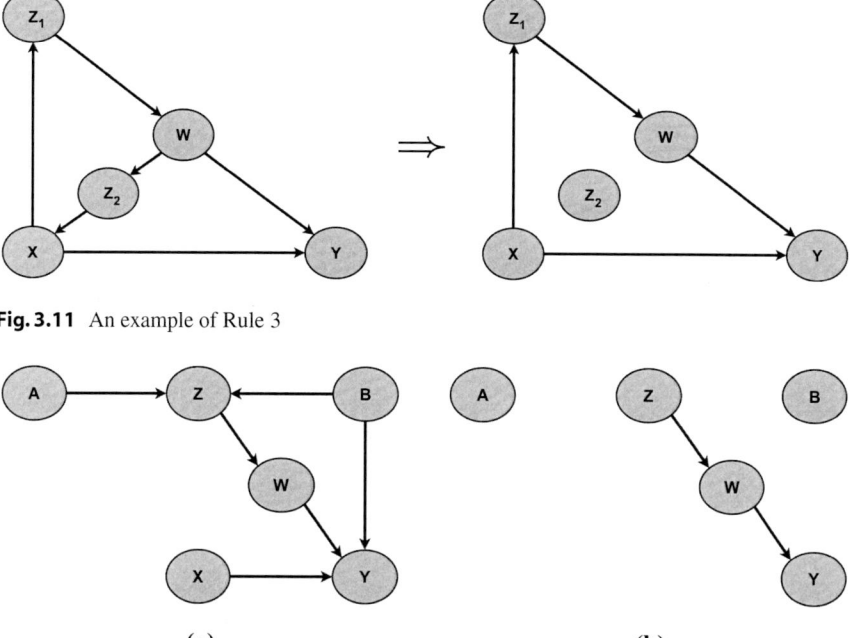

Fig. 3.11 An example of Rule 3

Fig. 3.12 Another example of Rule 3 that illustrates that the arrow from Z to W cannot be removed, as W affects Y

Rule 3 (Fig. 3.11) is the more complex one. It tells us when we can completely remove a *Do* expression rather than converting it to an observed quantity. This means that we can ignore an intervention if it doesn't influence the outcome through any uncontrolled path; we can remove $Do(z)$ if there is no causal association (or no unblocked causal paths) flowing from Z to Y. However, we should not delete the arrows into the Z nodes that are parents of W, why? This is because if we remove these arrows, it could alter the distribution of Y, removing possible paths from other variables that affect Y, see Fig. 3.12.

Applying the Do-Calculus is analogous to deriving a proof. That is, transformations to the initial causal query are performed by applying one of the three rules or a probability axiom. At the end, we obtain an expression based only on associational probabilities (observed variables) or we do not find a solution. Although this may seem too complex, fortunately there is a polynomial-time algorithm for the Do-Calculus [9].

3.4 Counterfactuals

Counterfactuals are a way of reasoning that we commonly perform in our life; that is, imagine scenarios that are different form what really happen. For instance, consider the causal model of Fig. 3.1a. A typical counterfactual question would be: A person did not exercise (*Lack of Exercise* = TRUE) and he suffered a stroke (*Stroke* = TRUE), would he not have suffered the stroke (*Stroke* = FALSE) if he had exercised more (*Lack of Exercise* = FALSE)?

Counterfactual reasoning is more complex than associational or interventional reasoning, as we will see later on. It involves analyzing alternative, conflicting scenarios; and requires the transfer of information from one scenario to the other. For this type of reasoning we have to go beyond causal Bayesian networks, as these models are not sufficient to solve counterfactual queries. We require that the causal relations are represented as functions, that is, *structural causal models* (SCM), also known as functional Bayesian networks.

In general, as we go up in the *ladder of causation*, we requiere more information. Reasoning about interventions requieres in addition to the probability distribution, the causal graphical model; and counterfactual reasoning requires, in addition to the causal structure, the ability to transfer information from on scenario (the actual or real one) to another scenario (the imaginary one). For this we require a structural (functional) causal model because using a causal Bayesian network we can loose relevant information required for counterfactual reasoning.[2] In particular, what was learned from the actual situation has to be transferred to the imaginary situation. This is done by keeping the probability distribution of the exogenous variables in a SCM from one scenario to another, which is not possible with a causal Bayesian network.

We will start by reminding the definition of a structural causal model (or causal model for short), and then describe how we can perform counterfactual reasoning for a deterministic model, and for a probabilistic model.

3.4.1 Structural Causal Models: A Review

Consider a causal model, $M(\mathbf{U}, \mathbf{V}, \mathbf{F})$, where:

- $\mathbf{U}$ is a set of exogenous variables that are determined by factors outside the model.
- $\mathbf{V}$ is a set $V_1, V_2, ..., V_n$ of variables, called endogenous, that are determined by variables in the model.
- $\mathbf{F}$ is a set of functions $f_1, f_2, ..., f_n$ such that each function is a mapping from $U_i \cup PA_i$ to V_i, such that PA_i are the set of endogenous variables that are direct causes of V_i. This is, each function is of the form $v_i = f_i(pa_i, u_i)$. The entire set of endogenous variables has a unique solution $\mathbf{V}(u)$.

---

[2] It is alway possible to transform a causal Bayesian network to a structural causal model but not viceversa.

We will consider that all variables are binary. Given certain variable, X, x represents the value *true* (or 1, *yes*, ...) and $\bar{x}$ the value *false* (or 0, *no*, ...).

A causal model, M, is associated to a directed causal graph, $G(M)$, in which each node corresponds to a variable and the directed edges point from members of PA_i and U_i toward V_i.

Given an intervention $Do(X = x)$ on a subset of variables, a submodel $M_x = (\mathbf{U}, \mathbf{V}, \mathbf{F_x})$ of the causal model M is obtained. F_x is transformed by deleting from F all functions corresponding to members of subset X and replacing them with the set of constant functions $X = x$. Submodels are useful for representing the effect of hypothetical changes implied by counterfactuals.

Now, we can define a counterfactual: What would have been the value of Y had been $X = x$? That is, given $X \subset V$ and $Y \subset V$, $Y_x(u) = y$ is the potential response of Y to $X = x$ in the situation u. This implies a hypothetical modification of the equations in the causal model; it modifies the actual course of history by enforcing the condition $X = x$ with the minimal changes in the model.

The previous definition extends to probabilistic models, $(M, P(u))$ in which a probability function $P(u)$ is defined over the domain of the exogenous variables,[3] U. The probability of a counterfactual is defined considering the submodel M_x as:

$$P(Y_x = y) = \sum_{u|Y_x(u)=y} P(u). \tag{3.13}$$

To compute the probability of the consequence of a contrafactual we can use the following procedure. Given a model $(M, P(u))$ and the evidence e:

1. Abduction—Update the probability $P(u)$ given the evidence to estimate $P(u \mid e)$.
2. Intervention—Modify the model M by the action $Do(x)$, where X is the antecedent of the counterfactual to obtain M_x.
3. Prediction—Use the modified model $(M_x, P(u \mid e))$ to compute the probability of y, $P(y_x \mid e)$, the consequence of the counterfactual.

3.4.2 Evaluating Counterfactuals

Next, we present examples of the evaluation of counterfactuals in a deterministic and a probabilistic case. Consider the following causal model (taken from [4]), see Fig. 3.13. A court (U) gives an execution order, so the captain (C) gives a signal to the firing squad that includes two riflemen, R_1 and R_2. When they receive the signal, both rifleman shoot and the prisoner dies (D).

---

[3] Note that in this type of models, the uncertainty is associated to the exogenous variables, and the functional equations are all deterministic.

Fig. 3.13 Causal graph for
the firing squad example.
U—court order, C—captain,
R_1 and R_2—riflemen,
D—prisoner

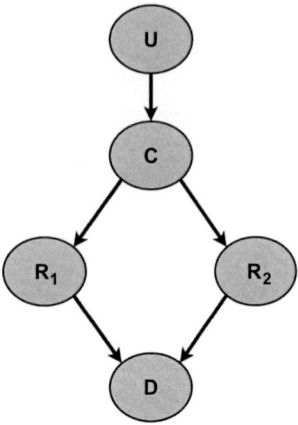

3.4.2.1 Deterministic Case

In the deterministic case, a functional causal model of the previous story is the
following:

$$C = U$$
$$R_1 = C$$
$$R_2 = C$$
$$D = R_1 \lor R_2$$

Lets consider de following counterfactual question: "The prisoner is dead, would
the prisoner be dead even if rifleman R_1 had not shot?". To evaluate the counterfactual
we modify the model according to the new fact (*intervention*) and we obtain the
submodel $M_{\overline{R_1}}$:

$$C = U$$
$$\overline{R_1}$$
$$R_2 = C$$
$$D = R_1 \lor R_2$$

Next, we do the *prediction* step according to the evidence, U and $\overline{R_1}$, and we
obtain the following conclusions[4]: C, R_2, D.

3.4.2.2 Probabilistic Case

Now we consider the same story but when there is uncertainty in the model:

- There is a probability $P(U) = p$ that the court orders the execution.
- There is a probability that R_1 is nervous, N, with $P(N) = q$, and fires without
receiving the order from the captain. N is independent of U.

---

[4] Note that in this case we do not apply the abduction step as there are no probabilities.

We want to answer the following counterfactual question: "If the prisoner is dead, what is the probability that the prisoner is alive if rifleman R_1 had not shot?"; that is, $P(\overline{D}_{\overline{R}_1} \mid D)$.

Now we have a probabilistic functional causal model, $(M, P(u, n))$, as follows:

$C = U$

$R_1 = C \vee N$

$R_2 = C$

$D = R_1 \vee R_2$

With the corresponding probability distribution:

$$P(u, n) = \begin{cases} pq & u = 1, n = 1 \\ p(1 - q) & u = 1, n = 0 \\ (1 - p)q & u = 0, n = 1 \\ (1 - p)(1 - q) & u = 0, n = 0 \end{cases}$$

Following the counterfactual evaluation procedure, we first apply the abduction step to evaluate $P(u, n \mid D)$. To evaluate it we consider two possibilities:

1. If not court order ($u = 0$) and the rifleman R_1 is not nervous ($n = 0$), then the probability es zero, $P(u, n \mid D) = 0$
2. Other wise, the prisoner will be death (D) if the court give the order ($u = 1$) or R_1 is nervous ($n = 1$), so $P(D) = 1 - (1 - p)(1 - q)$. Thus, $P(u, n \mid D) = P(u, n)/P(D) = P(u, n)/[1 - (1 - p)(1 - q)]$.

Next, we apply the intervention step generating the submodel $M_{\overline{R}_1}$:

$C = U$

$\overline{R_1}$

$R_2 = C$

$D = R_1 \vee R_2$

Keeping the same probability distribution $P(u, n \mid D)$.

Finally, we apply the prediction step to calculate $P(\overline{D})$. For the prisoner to be alive implies that the court did not give the order (although we assume that R_1 did not shot, if the court gives the order, the captain will signal and R_2 will shot); that is, $\overline{D} \Rightarrow \overline{U}$. So we can obtain $P(\overline{D}_{\overline{R}_1} \mid D)$ given the probabilistic model as:

$$P(\overline{D}_{\overline{R}_1} \mid D) = P(\overline{U} \mid D) = P(\overline{U})/P(D) = [(1-p)q]/[1-(1-p)(1-q)] \quad (3.14)$$

For the probabilistic case, next we present a more practical approach to evaluate counterfactuals based on a graphical representation.

3.4.3 Twin Network Approach

When evaluating a counterfactual it is required to keep the joint probability distribution of the exogenous variables from the *real* scenario to the *hypothetical* one. The previous approach, based on a probability table, results impractical when there are several exogenous variables, as it is required to keep their joint distribution. A

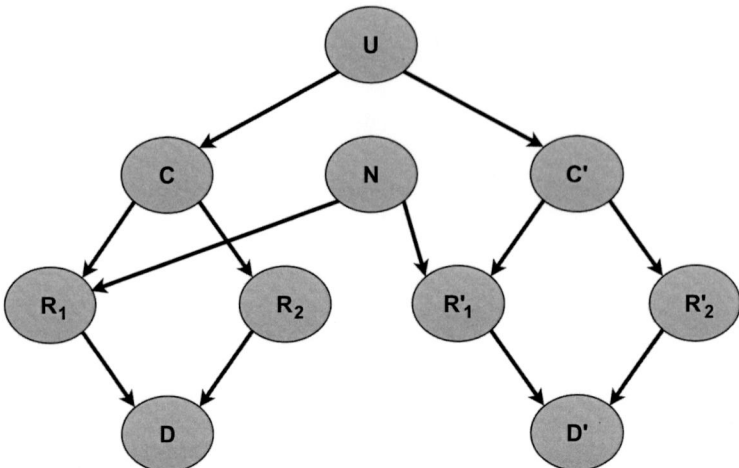

Fig. 3.14 Twin network for the firing squad example. The subgraph on the left side represents the *actual* scenario and the subgraph on the right side the *hypothetical* scenario. The exogenous variables, U and N, are common to both scenarios

practical way to do this is through the twin network technique. Basically, we duplicate the endogenous variables in the model but have only one copy of the exogenous variables.

The twin network graph for the firing squad example is depicted in Fig. 3.14. Note that both networks of endogenous variables are identical, and they share the exogenous (background) variables.

Given the twin network representation, to compute the previous counterfactual query: "If the prisoner is dead, what is the probability that the prisoner is alive if rifleman R_1 had not shot?"; that is, $P(\overline{D_{\overline{R_1}}} \mid D)$, we do the following. We instantiate D to true in the network (left part in the figure, corresponding to the real scenario). Then, we delete the arrows pointing to R_1' (right part of the figure corresponding to the hypothetical scenario), set R_1' to false and compute $P(\overline{D'_{\overline{R_1}}} \mid D)$.

3.4.4 Prototypical Counterfactual Queries

To conclude, we present a more realistic example of counterfactual reasoning and, based on this example, some prototypical counterfactual queries.[5] Consider a model that represents the relation between taking certain drug (X) and the probability of dying (Y). Figure 3.15 shows a simplified graphical model, where *Type* (U) represents the different possible reactions to the drug for different patients (a hidden variable).

---

[5] Example inspired on [2].

Fig. 3.15 An example of a causal Bayesian network to illustrate prototypical counterfactual queries. *Drug* and *Death* are observed variables, *Type* is unobserved

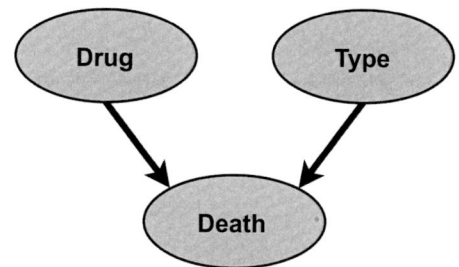

A prototypical query is: "A person was given the drug and died, what is the probability that she would be alive if she did not take the drug". This is known as the *probability of necessity*:

$$PN = P(\overline{y}_{\overline{x}} \mid x, y) \tag{3.15}$$

That is the probability she will not be death if she is not given the drug (imaginary scenario), given that she was given the drug and died (actual scenario). Situations like this are can be considered in a trial; for example, if $PN > 0.5$, could imply that the drug is harmful.

Another prototypical query is the *probability of sufficiency* –the probability that she will die if given the drug, given that she was not given the drug and she is alive:

$$PS = P(Y_x \mid \overline{y}, \overline{x}) \tag{3.16}$$

If we combine the two previous cases, we get the *probability of necessity and sufficiency*. That is, the joint probability of both scenarios:

$$PNS = P(x, y)PN + P(\overline{x}, \overline{y})PS \tag{3.17}$$

An interesting product of this last query is that we might recover the prior probabilities of the hidden variables in the model. Thus, counterfactuals reasoning provides information that it can not be obtained from associational or interventional reasoning (see [2,4]).

3.5   Additional Readings

The book by Pearl [4] is the main reference in this field, providing a more extensive introduction to causal reasoning. A text that provides a gentler introduction to causal reasoning is [5], and [6] is another book oriented to a wider audience. The lectures by A. Darwiche give an excellent introduction to predictive and counterfactual reasoning [2]. Peters et al. [7] provides a general introduction to causality, with emphasis on its relation with statistics and machine learning. An introduction to causal inference and discovery in Python is in [3].

3.6 Exercises

1. Given the CBN in Fig. 3.1a, and the following CPTs (HC-Health Consciousness, LE-Lack of Exercise, UD-Unhealthy diet, O-Overweight, S-Stroke):

$$P(LE \mid HC) = \begin{array}{c|cc} HC & F & T \\ \hline LE\ F & 0.2 & 0.7 \\ T & 0.8 & 0.3 \end{array}$$

$$P(UD \mid HC) = \begin{array}{c|cc} HC & F & T \\ \hline UD\ F & 0.4 & 0.9 \\ T & 0.6 & 0.1 \end{array}$$

$$P(O \mid LE, UD) = \begin{array}{c|cccc} LE, UD & F,F & F,T & T,F & T,T \\ \hline O\quad F & 0.8 & 0.6 & 0.7 & 0.2 \\ T & 0.2 & 0.4 & 0.3 & 0.8 \end{array}$$

$$P(S \mid O) = \begin{array}{c|cc} S & F & T \\ \hline O\ F & 0.8 & 0.6 \\ T & 0.2 & 0.4 \end{array}$$

   (a) Given the intervention $do(Unhealthy-diet = true)$, calculate the posterior probability of $Overweight$ and $Stroke$. (b) Repeat for $do(Unhealthy-diet = false)$.

2. Based on the results from the previous problem, what is the causal effect of $Unhealthy-diet$ on $Overweight$? And on $Stroke$? (use the same parameters as in Exercise 1).

3. Given the CBN in Fig. 3.1, what is the causal effect of $Stroke$ on $Unhealthy-diet$? And on $Lack-of-Exercise$? (Use the same parameters as for Exercise 1.)

4. Given the CBN in Fig. 3.16, list all the subsets of variables that satisfy the backdoor criterion for the causal relation $X \rightarrow Y$.

5. Given the CBN in Fig. 3.16, list all the subsets of variables that satisfy the frontdoor criterion for the causal relation $X \rightarrow Y$.

Fig. 3.16 A CBN

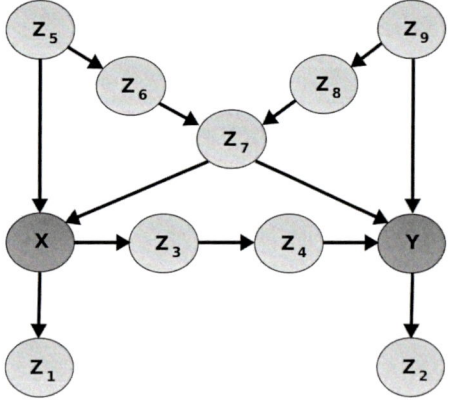

Fig. 3.17 Causal model for
the kidney stones problem

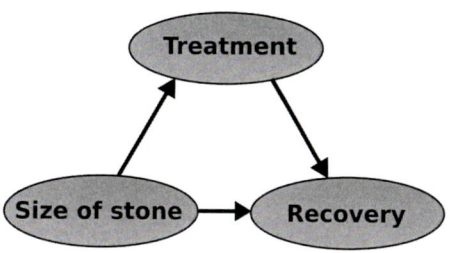

6. The table below shows data from a real medical study [1] comparing the success
   rates of two treatments for kidney stones. The table includes the success rate
   (Recovery, R) of the patients according to the Treatment (T) (A or B) and the
   Size of Stone (S). The numbers in parentheses indicate the number of success
   cases over the total size of the group. In total there are 700 patients, 350 were
   given treatment A and 350 treatment B.

Stone Size	Treatment A	Treatment B
Small stones	93% (81/87)	87% (234/270)
Large stones	73% (192/263)	69% (55/80)
Both	78% (273/350	83% (289/350)

   There is an apparent paradox, known as Simpon's Paradox, as the data seems to
   indicate that treatment A is better for each group, small and large stones, however
   considering both, treatment B seems better! A way to solve this paradox is via a
   causal Bayesian network, its structure is shown in Fig. 3.17. (a) Given the data,
   estimate the CPTs for the CBN. (b) Given this model, apply the do-calculus to
   obtain $P_{do(T=A)}(R = true)$ and $P_{do(T=B)}(R = true)$. (c) Which treatment is
   better, i.e. has a higher recovery rate?

7. For the firing squad example, consider the probabilistic case given in Sect. 3.4.2.2.
   Given $p = 0.8$ and $q = 0.1$, compute the following counterfactual query: "If
   the prisoner is dead, what is the probability that the prisoner is alive if rifleman
   R_1 had not shot?"

8. For the firing squad example, consider the probabilistic case given in Sect. 3.4.2.2.
   Given $p = 0.8$ and $q = 0.1$, compute the following counterfactual query: "If the
   prisoner is alive, what is the probability that the prisoner will be dead if rifleman
   R_2 had not shot?"

9. *** Implement the twin network model for the firing squad example, and verify
   the results of the previous two problems.

10. *** Develop a program that implements the the do-calculus for discrete causal
    Bayesian networks.

Acknowledgements Some sections of this chapter are inspired on [4]. Examples on prediction and identification of causal effects are taken from [2]. Figures 3.2 and 3.3 used with permission.

References

1. Charig CR, Webb DR, PayneS R, Wickham JE (1986) Comparison of treatment of renal calculi by open surgery, percutaneous nephrolithotomy, and extracorporeal shockwave lithotripsy. Br Med J 292(6524):879–882
2. Darwiche A (2023) Lectures on causality, course on learning and reasoning with bayesian networks, UCLA automated reasoning group, at https://web.cs.ucla.edu/~darwiche/
3. Molak A (2023) Causal inference and discovery in python. Packt Publishing, UK
4. Pearl J (2009) Causality: models, reasoning and inference. Cambridge University Press, New York
5. Pearl J, Glymour M, Jewell NP (2016) Causal inference in statistics: a primer. Wiley, UK
6. Pearl J, Mackienze D (2018) The book of why. Basic Books
7. Peters J, Janzing D, Scholkpf B (2017) Elements of causal inference. MIT Press
8. Shpitser I, Pearl J (2006) Identification of conditional interventional distributions. In: Proceedings of the twenty-second conference on uncertainty in artificial intelligence. AUAI Press, Corvallis, OR, pp 437–444
9. Shpitser I, Pearl J (2006) Identification of joint interventional distributions in recursive semi-Markovian causal models. AAAI, pp 1219–1226
10. Spirtes P, Glymour C, Scheines R (2000) Causation, prediction, and search. MIT Press

Part II
Algorithms

This part is dedicated to the main algorithms for causal discovery, including those based on observational and interventional data, as well as causal discovery from temporal data. It includes a chapter on the relation between causal models and reinforcement learning, addressing how to use causal knowledge in reinforcement learning, and how to simultaneously learn a causal model and an optimal policy during the exploration process.

Causal Discovery from Observational Data

4

Abstract

This chapter covers the topic of learning causal models from observational data. A general introduction to causal discovery is presented, highlighting some of the challenges. Then, the types of graphs used to represent partial models are introduced. Several algorithms for causal discovery from observational data are explained, including score-based and constraint-based methods, causal discovery with functional and parametric constraints, and techniques based on continuous optimization. To conclude the chapter, we address subject-specific causal discovery and a method to direct undirected edges based on causal effects estimation.

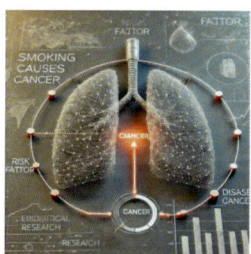

4.1 Introduction

Learning causal models from observational data (without direct interventions) poses many challenges. If we discover that there is a certain dependency between two variables, X and Y, we can not determine, without additional information, if X *causes* Y or vice versa. Additionally, there could be some other variable (cofactor) that produces the dependency between these two variables. For instance, consider that we

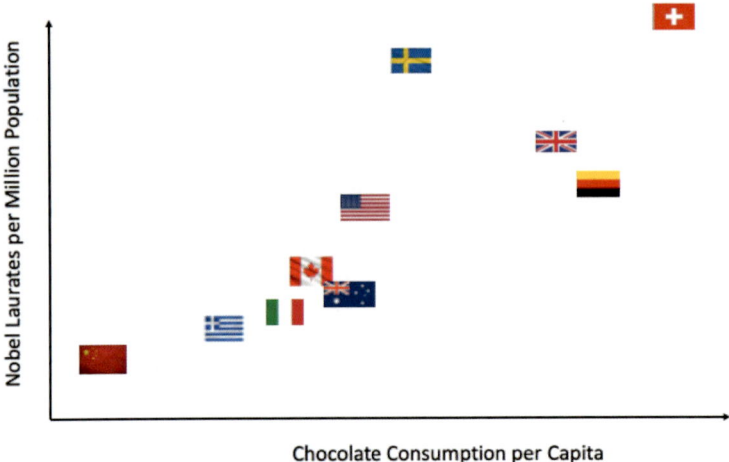

Fig. 4.1 Relation between annual chocolate consumption per capita and the number of Nobel prizes per million population for some countries. Figure based on [11]

have the following graph that represents the relation between chocolate consumption and number of Noble laureates from several countries, see Fig. 4.1. Based on this data, we observe that countries with higher chocolate consumption tend to have more Nobel prizes. Then, we might be inclined to conclude that eating chocolate increases the probability of a Nobel prize (Fig. 4.2a). However, there might be another variable that produces this apparent causal relation, known as a *latent common cause*. It could be that both, chocolate consumption and number of Nobel laureates, have to do with the income level of the country, as countries with high income levels tend to consume more chocolate and at the same time have more Nobel prizes (see Fig. 4.2b). Thus, a difficulty of learning causal relations lies in how to include in the model all the relevant factors.

Reichenbach [16] established a relation between graphical causal models and dependence known as the *common cause principle*: If two random variables X and Y are statistically dependent, then there exists a third variable Z that causally influences both. As a special case, Z may coincide with either X or Y. This principle is illustrated in Fig. 4.3.

In general, it is not possible to obtain a unique structure for a causal Bayesian network given only observational data; we obtain what is known as a *Markov equivalence class* (MEC). For instance, if we consider three variables, and the following *skeleton*: $X - Y - Z$, there are four possible directed graphs, as shown in Fig. 4.4. Three of these graphs represent the same conditional independence relation, X and Z are independent given Y, so they correspond to a Markov equivalence class. The other one is a different MEC, in which X and Z are not independent given Y.

A Markov equivalence class includes all the graphs that entail the same set of D-separations –same conditional independences under the *faithfulness* assumption (described below). All the graphs in a MEC satisfy the following two properties:

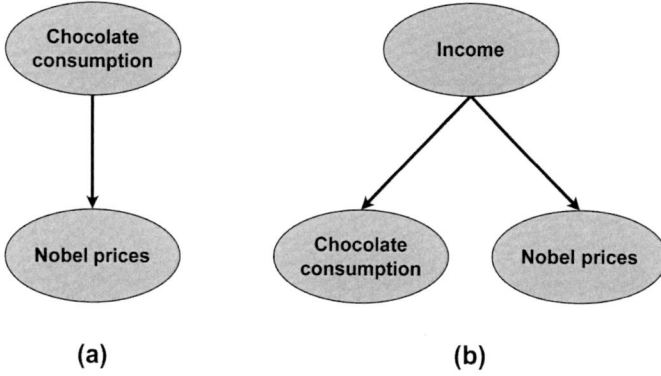

Fig. 4.2 An example of the difficulty of learning causal models. **a** The initial causal model that shows an apparent causal relation between "Chocolate consumption" and "Nobel prices". **b** An alternative causal graph with a common cause, "Income", that explains the dependency between "Chocolate consumption" and "Nobel prices"

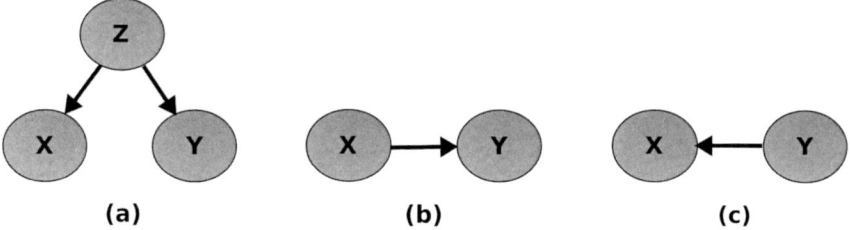

Fig. 4.3 Common cause principle. A statistical dependence between X and Y indicates that both are caused by a third variable: **a** Z causes both, X, Y, or **b** X causes Y, or **c** Y causes X. It is also possible that there is a direct causal relation between X and Y and a cofactor Z

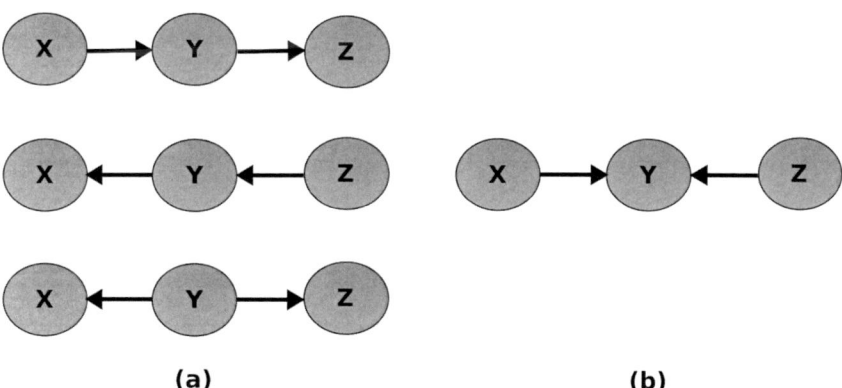

Fig. 4.4 Dependence graphs for three variables with the skeleton $X - Y - Z$. **a** These graphs represent that X and Z are conditionally independent given Y. **b** This graph corresponds to the case in which X and Z are not independent given Y

1. They have the same skeleton, that is, the same underlying undirected graph (if all directed edges are made undirected).
2. They have the same V structures, that is, subgraphs of the form $X \rightarrow Y \leftarrow Z$, such that there is no arc between X and Z. This substructures are also known as *colliders*.

If causal sufficiency is not assumed, then what we can learn are *maximal ancestral graphs* (MAGs), and the set of equivalent MAGs known as *partial ancestral graphs*. In the following section, we discuss in detail the different types of graphs.

To obtain the structure of a causal model from observational data, additional assumptions are required. In general, the following assumptions are used when learning the structure of causal networks:

Causal Markov Condition: a variable is independent of its non-descendants given its direct causes (parents in the graph).

Faithfulness: there are no additional independencies between the variables in the model that are not implied by the causal Markov condition.

Causal Sufficiency: there are no common unobserved confounders (latent or hidden variables) of the observed variables in the model.

Other assumptions can be made, such as to consider a particular type of relations between a variable and it causes (e.g., linear models) and/or particular distributions about the uncertainty/noise (e.g., Gaussian noise).

Next, we present the different types of graphs used when learning causal models; and later the main types of causal discovery algorithms.

4.2   Types of Graphs

A causal Bayesian network (CBN) is represented as a directed acyclic graph or DAG. CBNs include only directed arcs in the graph that represent *causal* relations; such as $A \rightarrow B$. As for Bayesian networks, the conditional independence relations can be read directly from the graph via the criterion known as *D-separation*.

Causal discovery algorithms can not always obtain a unique causal model, that is a DAG, so other types of graphs are used to represent the (partial) structure obtained. Next we introduce the representations used when assuming, or not, causal sufficiency.

4.2.1   Markov Equivalence Classes Under Causal Sufficiency

Under causal sufficiency, in general, what we can learn from observational data is a Markov equivalence class.

Several DAGs can encode the same conditional independencies via d-separation. Such DAGs form a **Markov Equivalence Class (MEC)** which can be described

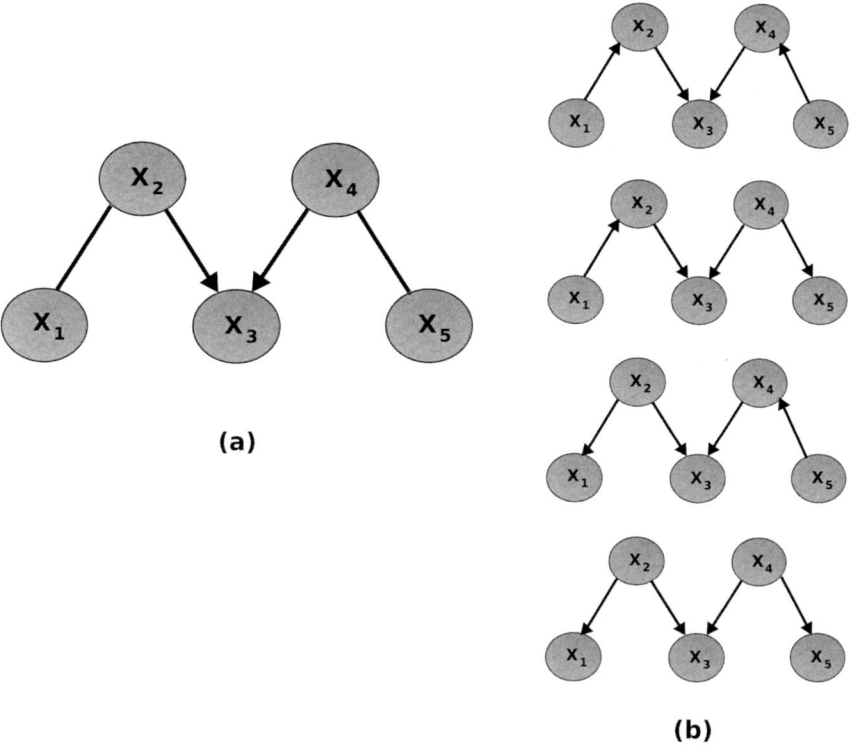

Fig. 4.5 a An example of a CPDAG, and **b** the set of DAGs it encodes. Note that in all DAGs, $I(X_1, X_2, X_3)$, $I(X_3, X_4, X_5)$ and $\cancel{I}(X_2, X_3, X_4)$

uniquely by a **Completed Partially Directed Acyclic Graph (CPDAG)**. A CPDAG $\mathscr{C}$ has the same adjacencies as any DAG in the MEC encoded by $\mathscr{C}$. A directed edge $X \to Y$ in a CPDAG $\mathscr{C}$ corresponds to a directed edge $X \to Y$ in every DAG in the MEC described by $\mathscr{C}$. For any non-directed edge $X - Y$ in a CPDAG $\mathscr{C}$, the Markov equivalence class described by $\mathscr{C}$ contains a DAG with $X \to Y$ and a DAG with $X \leftarrow Y$. Thus, CPDAGs contain directed and non-directed edges. An example of a MEC and the corresponding DAGs is depicted in Fig. 4.5.[1] Note that all DAGs in the same Markov equivalence class have the same skeleton (undirected graph obtained if we make all directed edges undirected) and V structures (that is, subgraphs of the form $X \to Y \leftarrow Z$).

---

[1] Recall the $I(X_1, X_2, X_3)$ means that X_1 is independent of X_3 given X_2.

4.2.2   Markov Equivalence Classes with Unmeasured Variables

If some variables are unmeasured, and they can be common causes of the measured variables (cofactors), an alternative representation that takes into account these cofactors is required. We assume a set of variables $\mathbf{V}$, in which $\mathbf{O}$ are the subset of measured or observed variables and $\mathbf{L}$ the unmeasured or latent variables, such that $\mathbf{V} = \mathbf{O} \cup \mathbf{L}$.

Maximal Ancestral Graphs (MAGs) can represent conditional independence information and causal relationships in DAGs that include unmeasured (hidden or latent) variables. A MAG represents a DAG after all latent variables have been marginalized out, and it preserves all entailed conditional independence relations among the measured variables. This is, given a DAG D over $\mathbf{V} = \mathbf{O} \cup \mathbf{L}$ there is a MAG M over $\mathbf{O}$ alone, such that for any disjoint sets $\mathbf{X}, \mathbf{Y}, \mathbf{Z} \subseteq \mathbf{O}$, $\mathbf{X}$ and $\mathbf{Y}$ are d-separated by $\mathbf{Z}$ in D if and only if they are *m-separated* by $\mathbf{Z}$ in the MAG M. We can define m-separation in a similar way to d-separation:

m-separation [20]: A path p between X and Y is active relative to $\mathbf{Z}$, $(X, Y \notin \mathbf{Z})$, if every non-collider on p is not in $\mathbf{Z}$, and every collider on p is an ancestor of some $Z \in \mathbf{Z}$. X and Y are m-separated by Z if there is no active path between X and Y relative to $\mathbf{Z}$.

The following construction gives us such a MAG: (i) For each pair of variables $X, Y \in \mathbf{O}$, X and Y are adjacent in M if and only if there is an *inducing path*[2] between them relative to $\mathbf{L}$ in D. (ii) For each pair of adjacent variables X, Y in M: (a) orient the edges $X \rightarrow Y$ in M if X is an ancestor of Y in D; (b) orient it as $X \leftarrow Y$ in M if Y is an ancestor of X in D; (c) orient it as $X \leftrightarrow Y$ in M, otherwise. An example of a DAG with latent variables and the corresponding MAG is shown in Fig. 4.6.

Several MAGs can also encode the same conditional independencies via m-separation. Such MAGs form a Markov Equivalence Class (MEC) which can be described uniquely by a *Partial Ancestral Graph* (PAG). A PAG, P_M, si a partial

Fig. 4.6 An example of a DAG (**a**) where U_1, U_2, U_3 are latent variables, and the corresponding MAG (**b**)

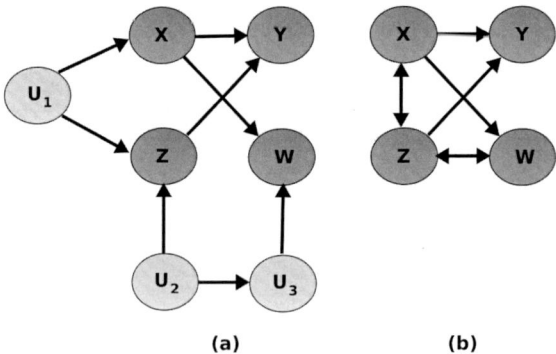

(a)   (b)

---

[2] See Sect. 4.3.4.3 for a formal definition of inducing path.

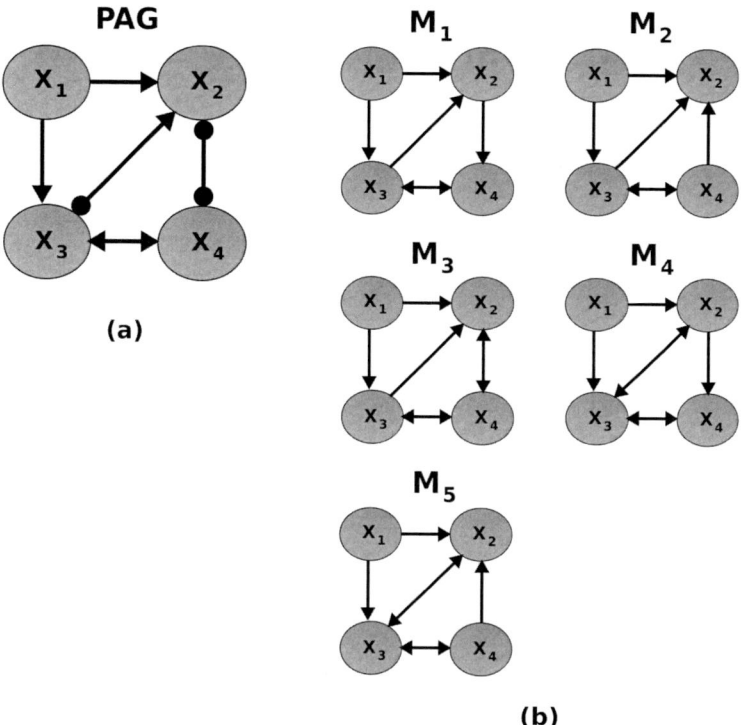

Fig. 4.7 **a** An example of a PAG; **b** the corresponding MAGs ($M_1, \ldots, M_5$) in the Markov equivalence class

mixed graph with three kinds of marks in its edges: arrowhead ($>$), tail ($-$) or circle ($\circ$), such that: (i) P_M has the same adjacencies as all the MAGs in the MEC; (ii) Every non-circle mark in P_M is an invariant mark in all MAGs in the MEC; i.e., a mark of arrowhead is in P_M if and only if it is shared by all MAGs in the MEC, and a mark of tail is in P_M if and only if it is shared by all MAGs in the MEC; (iii) A mark of circle otherwise (when the same mark is not shared by all MAGs).

Figure 4.7 depicts a PAG and the set of MAGs in the corresponding MEC.

4.3 Causal Discovery Algorithms

The type of datasets used in the learning of the causal structure may be: (i) **Observational**, data corresponding to measurements made under natural conditions of a causal system; (ii) **Experimental**, data correspond to measurements made under external interventions; (iii) **Hybrid**, it combines observational and experimental data. Ideally, experimental data should be used in the structure learning of causal BNs. Nevertheless, experimental data are not always available or it can be unethical, infea-

Table 4.1 Types of causal discovery algorithms

Dimension		
Sufficiency	Causal sufficiency	Unmeasured cofounders
Types of variables	Discrete	Continuous (linear Gaussian)
Type of algorithm	Constraint-based	Score-based
Restrictions	No restrictions	Parametric and/or Functional
Optimization	Combinatorial	Continuous

sible, time consuming, or expensive to obtain. On the other hand, observational data, i.e., data associated with processes that cannot be reproduced, are often abundant. In this chapter, we focus on learning from observational data, and in the next one from experimental data.

We can divide the problem of causal discovery from observational data according to three dimensions [23]: (i) If the data is independent and identically (i.i.d.) distributed or not, (ii) If we consider parametric constraints or not, (iii) If there are or not latent (hidden) cofounders. Depending on these dimensions, we can obtain a unique model, different types of equivalence classes, or other restricted sets of causal models. In this chapter we will focus on i.i.d. (static) data; for which, if there are not parametric constraints we can learn an equivalence class, and with certain parametric constraints we can recover a unique causal structure. In Chap. 6 we consider the case of non-independent, identically distributed (time-series) data, for which it is possible, under certain assumptions, to learn a unique causal structure.

For the i.i.d case, causal structure learning approaches can be classified according to several dimensions, as shown in Table 4.1.

In general, these methods recover a Markov equivalence class unless extra assumptions are made or they incorporate background knowledge. For multinomial (discrete) distributions and linear Gaussian functional equations, we can only discover from observational data a Markov equivalence class. However, for other types of models, such as linear non-Gaussian functional equations, it is possible to obtain a unique DAG, as we will see later.

Once we discover the *essential graph* (skeleton plus V structures) of a causal model, it is possible to derive the directions of some other arcs by a set of rules known as Meek rules.

There is a great variety of causal discovery algorithms (see [6, 22]). Here we present some representative examples of the different types of methods:

- score-based causal discovery,
- constraint-based causal discovery,
- causal discovery with functional and parametric constraints,
- causal discovery based on continuos optimization.

Before we discuss the causal discovery methods, we present the Meek rules and review some evaluation metrics. And after, we consider how to solve some of the

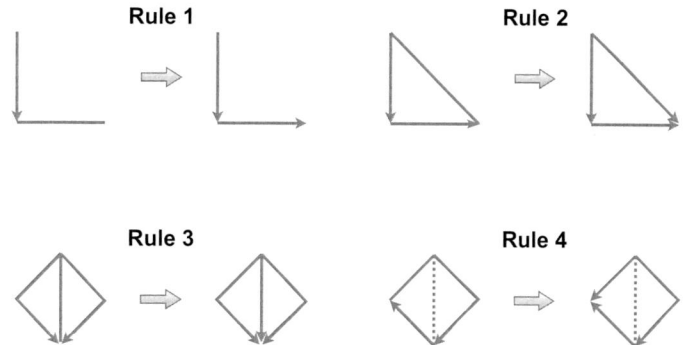

Fig. 4.8 The figure shows the four rules (known as Meek rules) for orienting some additional edges in a partially oriented causal graph

challenges: (i) small data sets, in particular for learning subject-specific causal models, and (ii) how to orient undefined edges.

4.3.1   Meek Rules

Once we have obtained a MEC, we can orient some other arcs in the graph given the already oriented arcs that correspond to the V structures in the partially oriented DAG. Meek [10] proposes a set of orientation rules that are based on the following principles: (i) do not generate additional unshielded V structures,[3] and (ii) do not generate directed cycles. The four orientation rules are easier to understand graphically, and are shown in Fig. 4.8. Each rule shows an initial partial subgraph (left) and the resulting partial graph after applying the rule (right). The dashed line in Rule 4 means that this arc can be undirected or directed in any direction ($-$, $\leftarrow$, $\rightarrow$). These rules are sound and complete [10].

4.3.2   Evaluation Metrics

There are several evaluation metrics to evaluate the performance of causal discovery algorithms, here we focus on those used to evaluate the structure[4]:

True Positive Rate (TPR):   relative number of discovered edges along the ground truth edges with a probability above certain threshold $t \in (0, 1)$.

---

[3] A V estructure is unshielded if there is no arc between the variables at the tails of the arcs.
[4] Other measures can be used to evaluate the parameters of the model, we will discuss these when we cover score-based methods.

False Positive Rate (FPR): relative number of discovered edges along the ground
truth missing edges with a probability above certain threshold $t \in (0, 1)$.

Area Under the Curve (AUC): area under the curve of (FPR_t, TPR_t) where the
threshold, t, is varied between 0 and 1.

Structural Hamming Distance (SHD): number of required changes to the graph for
it to match the ground truth, including missing edges, extra edges, and incorrect
edges.

4.3.3 Score-Based Causal Discovery

Score-based approaches test the validity of a candidate causal graph, G, according
to some scoring function S. The objective is to find the graph G^* that maximizes a
score based on the empirical data, D:

$$G^* = argmax_G S(D, G) \qquad (4.1)$$

There are several scoring function such as Bayesian Information Criterion (BIC),
the Minimum Description Length (MDL), the Bayesian Gaussian equivalent (BGe)
score, the Bayesian Dirichlet equivalence (BDe) score, the Bayesian Dirichlet equivalence uniform (BDeu) score, and others.

As the number of potential graphs is very large (super-exponential on N, the
number of variables; for example the number of possible DAGs for 10 variables
is greater than 4×10^{18}), score-based methods perform a heuristic search over the
possible graphs, usually starting form a *simple graph* (i.e., a tree) and doing local
changes in the structure (add, delete, or invert an arc), until the score can not be longer
improved. Different search strategies are used, from simple ones as hill-climbing, to
sophisticated optimization techniques such as genetic algorithms. These methods in
general obtain a local optimum, as it difficult to guarantee a global maximum. They
are affected by the quantity and quality of the data to estimate the score.

Next we present some scoring functions; and then we describe two score-based
methods: K2, a classical algorithm that can be applied to causal discovery if the
causal order of the variables is known; and Greedy Equivalence Search, a well known
score-based causal discovery algorithm that returns a CPDAG.

4.3.3.1 Scoring Functions

There are several possible fitness measures or scoring functions. Two desirable properties for scoring functions are:

Decomposability: a scoring function, S, is decomposable if it can be expressed as
the product of local functions that only depend of a node X_i and its parents $Pa(X_i)$:
$S = \prod_i f(X_i, Pa(X_i), \mathbf{D})$; where $\mathbf{D}$ is the data set over which is evaluated the
causal structure. This is important for efficiency reasons during the search process;

given this property when a local change is made to the structure, only a part of the score has to be re-evaluated.

Score Equivalence: scoring function, S, is score equivalent if for any pair of equivalent DAGs, G and G', it assigns the same score, $S(G) = S(G')$. Since the scoring function is score equivalent, any DAG contained in the candidate MEC could be used for evaluating that MEC.

Next we describe some common scoring functions, including: the maximum likelihood (ML), the Bayesian information criterion (BIC), the Bayesian score (BD), and the minimum description length (MDL) criterion.

The maximum likelihood score selects the structure that maximizes the probability of the data, D, given the structure, G:

$$G^* = ArgMax_G[P(D \mid \Theta_G, G_i)] \qquad (4.2)$$

where G_i is the candidate structure and Θ_G the corresponding vector of parameters (probability of each variable given its parents according to the structure).

The direct application of the ML score might result in a highly complex network, which usually implies overfitting the data (poor generalization) and also makes inference more complex. Therefore, a way to penalize complex models is required.

A commonly used scoring function that includes a penalty term is the Bayesian Information Criterion or BIC defined as:

$$BIC = log P(D \mid \Theta_G, G_i) - \frac{d}{2} log N \qquad (4.3)$$

where d is the number of parameters in the BN and N the number of cases in the data. An advantage of this metric is that it does not requiere a prior probability specification and it is related to the MDL measure, compromising between the precision and complexity of the model. However, given the high penalty on the complexity of the model, it tends to choose structures that are too simple.

Bayesian scores

An alternative metric is obtained by following a Bayesian approach, obtaining the posterior probability of the structure given the data with the Bayes rule:

$$P(G_i \mid \mathbf{D}) = P(G_i)P(\mathbf{D} \mid G_i)/P(\mathbf{D}) \qquad (4.4)$$

Given that $P(\mathbf{D})$ is a constant that does not depend on the structure, it can be discarded from the metric to obtain the Bayesian or BD score:

$$BD = P(G_i)P(\mathbf{D} \mid G_i) \qquad (4.5)$$

$P(G_i)$ is the prior probability of the model. This can be specified by an expert or defined such that simpler structures are preferred; or just set to a uniform distribution.

The BDeU score is a variation of the BD score which makes the following assumptions: (i) the parameters are independent and have a prior Dirichlet distribution, (ii) equivalent structures have the same score, (iii) the data samples are independent and identically distributed (iid). Under these assumptions the *virtual counts* required to compute the score can be estimated as:

$$N_{ijk} = P(X_i = k, Pa(X_i) = j \mid G_i, \Theta_G) \times N' \qquad (4.6)$$

This is the estimated count of a certain *configuration*: $X_i = k$ given $Pa(X_i) = j$; N' is the equivalent sample size.

The *Bayesian Dirichlet equivalent and Uniform* (BDeU) is a decomposable and score equivalent function that evaluates MECs defined over discrete variables with complete data sets $\mathbf{D}$ (without missing values) [8]:

$$BDeU(\mathscr{G}, \mathbf{D}) = \prod_{i=1}^{n} \{score(X_i, \mathbf{Pa}(X_i), \mathbf{D})\} \qquad (4.7)$$

$$score(X_i, \mathbf{Pa}(X_i), \mathbf{D}) = \prod_{j=1}^{q_i} \frac{\Gamma(\alpha_{ij})}{\Gamma(\alpha_{ij} + C_{ij})} \prod_{k=1}^{r_i} \frac{\Gamma(\alpha_{ijk} + C_{ijk})}{\Gamma(\alpha_{ijk})} \qquad (4.8)$$

Where n is the number of nodes in G, q_i is the number of values of $\mathbf{Pa}(X_i)$, r_i is the number of values of X_i, C_{ijk} is the number of cases in which $X_i = k$ and its parents $\mathbf{pa}(X_i = k) = j$, $N_{ij} = \sum_k C_{ijk}$, and $\alpha_{ijk} = \frac{1}{r_i q_i}$ is a Dirichlet prior parameter with $\alpha_{ij} = \sum_k \alpha_{ijk}$.

By assuming that the hyper parameters of the priors are one, we can further simplify the calculation of the Bayesian score, and obtain what is known as the K2 metric.[5] This score is decomposable and it is calculated for each variable X_i given its parents $Pa(X_i)$:

$$S_i = \prod_{j=1}^{q_i} \frac{(r_i - 1)!}{(N_{ij} + r_i - 1)!} \prod_{k=1}^{r_i} \alpha_{ijk}! \qquad (4.9)$$

where r_i and q_i as defined above, α_{ijk} is the number of cases in the database where $X_i = k$ and $Pa(X_i) = j$, and N_{ij} is the number of cases in the database where $Pa(X_i) = j$.

This metric provides a practical alternative for evaluating a CBN. Another common alternative that is based on the MDL principle is described next.

MDL

The MDL (*Minimum Description Length*) measure makes a compromise between accuracy and model complexity. Accuracy is estimated by measuring the mutual information between each variable and its parents. Model complexity is evaluated by counting the number of parameters. A constant, α within [0, 1], is used to balance

---

[5] The K2 algorithm is described next.

the weight of each aspect, that is, accuracy against complexity. The fitness measure is given by the following equation:

$$MC = \alpha(W/Wmax) + (1 - \alpha)(1 - L/Lmax) \qquad (4.10)$$

where W represents the accuracy of the model, and L the complexity. $Wmax$ and $Lmax$ represent the maximum accuracy and complexity, respectively. To determine the maximums, usually an upper bound is set on the number of parents each node is allowed to have. A value of $\alpha = 0.5$ gives equal importance to the model complexity and accuracy, while a value near 0 gives more importance to the complexity, and a value near 1 more importance to accuracy.

Complexity is given by the number of parameters required for representing the model, which can be measured with the following equation:

$$L = S[klog_2 n + d(S - 1)F] \qquad (4.11)$$

where n is the number of variables in the model, k is the average number of parents per variable, S is the average number of values per variable, F is the average number of values per parent variable, and d the number of bits per parameter. For example, consider a CBN with 16 variables, all variables are binary and have in average 3 parents, and each parameter is represented by 16 bits. Then:

$$L = 2 \times [3 \times log_2(16) + 16 \times (2 - 1) \times 2] = 2 \times [12 + 32] = 88$$

The accuracy can be estimated based on the 'weight' of each node. The weight of each node, X_i, is estimated based on its mutual information with its parents, $Pa(X_i)$:

$$w(X_i, Pa(X_i)) = \sum_{xi} P(X_i, Pa(X_i))log[P(X_i, Pa(X_i))/P(X_i)P(Pa(X_i))]$$

$$(4.12)$$

and the total weight (accuracy) is given by the sum of the weights for each node:

$$W = \sum_i w(X_i, Pa(X_i)) \qquad (4.13)$$

4.3.3.2   The K2 Algorithm

Given a *causal ordering*[6] for all the variables, learning the best structure is equivalent to selecting the best set of parents for each node independently. Initially, each variables has no parents. Then, the K2 algorithm incrementally adds parents to each node, as long as it increases the global score. When adding parents to any node does not increase the score, the search stops. Also, given a causal ordering it guarantees that there are no cycles in the graph.

---

[6] A causal ordering, $X_1, X_2, ...X_n$, implies that there could not be arcs from variables which have a higher number to variables with a lower number according to the order.

Algorithm 4.1 provides a summary of the K2 procedure. The inputs to the algorithm are the set of n variables with a causal ordering, $X_1, X_2, ...X_n$, a data base **D** containing m cases, and, usually, a restriction on the maximum number of parents for each variable, u. The output is the set of parents, $Pa(X_i)$, for each variable, which defines the structure of the network. Starting from the first variable according to the ordering, the algorithm tests all possible parents of a variable that have not been added, and includes the one that makes the maximum increment in the score of the network. This is repeated until there is no additional parent that increases the score; for every node on the network.

Algorithm 4.1 The K2 Algorithm

Require: Set of variables X with a causal ordering, scoring function S, and maximum parents u
Ensure: Set of parents for each variable, $Pa(X_i)$
  for $i = 1$ **to** n **do**

    $oldScore = S(i, Pa(X_i))$
    $incrementScore = true$
    $Pa(X_i) = \emptyset$
    while $incrementScore$ and $|Pa(X_i)| < u$ **do**

      let Z be the node in $Predecessors(X_i) - Pa(X_i)$ that maximizes S
      $newScore = S(i, Pa(X_i) \cup Z)$
      if $newScore > oldScore$ **then**

        $oldScore = newScore$
        $Pa(X_i) = Pa(X_i) \cup Z$

      else

        $incrementScore = false$

      end if

    end while

  end for
  return $Pa(X_1), Pa(X_2)...Pa(X_n)$

4.3.3.3 Greedy Equivalence Search

Greedy Equivalence Search (GES) [1,2] is a score-based (global) algorithm for causal structure learning, that assumes causal sufficiency and discrete variables. GES heuristically searches the structure in the space of Markov equivalence classes (MECs) in two stages. In each step of the algorithm, every candidate MEC is evaluated, and the MEC with the highest score that improves the score function is selected. In the first stage, starting with an empty graph, GES adds edges to candidate MECs until a local maximum is reached. In the second stage, it removes edges of the MEC found in the first stage; assuming that some *extra* edges might have been added in the first stage given the greedy nature of the search process. The algorithm stops when a local maximum is reached and returns the CPDAG that represents the found MEC. In the large sample limit, GES returns the Markov equivalence class containing the true causal graph. GES utilizes the $BDeU$ score.

Two extensions of GES are Fast GES (FGES) [15], and Greedy Fast Causal Inference (GFCI) [13]. GES has problems in discovering high-dimensional models. Fast Greedy equivalence search (FGES), a modification of GES, overcomes this limitation by optimizing and performing parallel score estimations. Considering that adding or deleting single edges impacts the set of parents of only a variable, FGES keeps an ordered list of the positive score differences between each variable and its old and new parents. In addition, FGES computes the score for a variable in parallel, with all possible single edge modifications. Further, to speed up the search of MECs with continuous variables, FGES consider a weak form of faithfulness: all edges $X \rightarrow Y$ for which X is uncorrelated to Y are not added. Greedy and Fast Causal Inference (GFCI), which does not require causal sufficiency, combines FGES and FCI (described in Sect. 4.3.4.3). First, GFCI applies FGES to find a Markov equivalence class. The edges in the MEC are undirected in the next step. After that, it uses FCI to remove false edges and correct the orientation of those edges in the MEC found by FGES.

4.3.4   Constraint-Based Causal Discovery

Constraint-based approaches test for conditional independencies in the joint distribution according to the empirical data, in order to construct a graph that reflects these conditional independencies. As mentioned before, there are often multiple graphs that fulfill a given set of conditional independencies, and so it is common for constraint-based approaches to output a graph representing some MEC. A challenge for this type of methods is that conditional independence tests require large sample sizes to be reliable.

Different independence tests can be used, for example the conditional *cross entropy*:

$$H(X, Y \mid \mathbf{Z}) = \sum_{X} \sum_{Y} P(X, Y \mid \mathbf{Z}) log_2 [P(X, Y \mid \mathbf{Z}) / P(X \mid \mathbf{Z}) P(Y \mid \mathbf{Z})]$$

(4.14)

The conditional cross entropy provides a measure of the mutual information (dependency) between two random variables, X, Y, conditioned on a set of variables $\mathbf{Z}$ (that could be empty); it is zero when the two variables are independent given $\mathbf{Z}$.

Next we describe PC, a popular constraint-based causal discovery algorithm, that assumes causal sufficiency. Then we present two alternative approaches that do not assume causal sufficiency, Bayesian Constraint-Based Causal Discovery (BCCD) and Fast Casual Inference (FCI).

4.3.4.1 PC Algorithm

The PC algorithm is probably the most well-known constraint-based structure learning algorithm. The PC-algorithm [20] proceeds in two stages, the first stage searches for adjacencies between variables to obtain the *skeleton* (underlying undirected graph), while in the second stage it finds the orientations of the edges, where possible.

To determine the skeleton, it starts from a fully connected undirected graph, and determines the conditional independence of each pair of variables given some subset of the other variables. For this, it assumes that there is a procedure that can determine if two variables, X, Y, are independent given a subset of variables, S, that is, $I(X, Y \mid S)$. An alternative for this procedure is the conditional cross entropy measure. If this measure is below a threshold value set according to a certain confidence level, the edge between the pair of variables is eliminated. These tests are iterated for all pairs of variables in the graph.

In the second phase, the direction of the edges are set based on conditional independence tests between variable triplets. It proceeds by looking for substructures in the graph of the form $X - Z - Y$ such that there is no edge $X - Y$. If X, Y are not

---

Algorithm 4.2 The PC algorithm

Require: Set of variables **X**, Independence test I
Ensure: Directed Acyclic Graph G
1: Initialize a complete undirected graph G'
2: i=0
3: **repeat**

4:      **for** $X \in \mathbf{X}$ **do**

5:          **for** $Y \in ADJ(X)$ **do**

6:              **for** $S \subseteq ADJ(X) - \{Y\}, \mid S \mid = i$ **do**

7:                  **if** $I(X, Y \mid S)$ **then**

8:                      Remove the edge $X - Y$ from G'

9:                  **end if**

10:              **end for**

11:          **end for**

12:      **end for**
13:      i=i + 1

14: **until** $\mid ADJ(X) \mid \leq i, \forall X$
15: Orient edges in G'
16: Return G

---

independent given Z, it orients the edges creating a V-structure $X \rightarrow Z \leftarrow Y$. Once all the V-structures are found, it attempts to orient the other edges based on Meek rules. Algorithm 4.2 summarizes the basic procedure.[7]

If the set of independencies are faithful to a graph[8] and the independence tests are perfect, the algorithm produces a Markov equivalence class the includes the original one; that is, the graph structure that generated the data.

The independence test techniques rely on having *enough* data for obtaining good estimates from the independence tests. Search and score algorithms are more robust with respect to the size of the data set, however their performance is also affected by the size and quality of the available data.

The PC algorithm assumes causal sufficiency, and an independence oracle that supplies the independence constraints true in the distribution that generated the data; if so, it recovers as much information about the true causal structure as possible with the available independence tests. However, it can not provide guarantees for a limited sample size. In general PC returns a CPDAG that represents the MEC, as there is no guarantee that the second phase of the algorithm can find the orientation of all the edges in the causal graph.

The results of the original PC algorithm depend on the variables' order according to which the independence tests are performed. However, there is a variant of this algorithm that makes it independent of the order [4].

Next, we will cover two extensions of the PC algorithm that relax the causal sufficiency assumption.

4.3.4.2 Bayesian Constraint-Based Causal Discovery

The Bayesian Constraint-Based Causal Discovery (BCCD) [3] algorithm is an extension of the PC algorithm that consists of two main phases:

1. Start with a completely connected graph and estimate the reliability of each causal link, $X - Y$, by measuring the conditional independence between X and Y. If a pair of variables are conditionally independent with a reliability above a certain threshold, it then deletes the edge between these variables.
2. The remaining causal relations (undirected edges in the graph) are ordered according to their reliability. Then, the edges in the graph are oriented starting from the most reliable relations, based on the conditional independence test for variable triplets.

To estimate the reliability of a causal relation, $R = X \rightarrow Y$, the algorithm uses a Bayesian score:

$$P(R \mid D) = \frac{P(D \mid M_R)P(M_R)}{P(D \mid M)P(M)} \quad (4.15)$$

---

[7] $ADJ(X)$ is the set of nodes adjacent to X in the graph.
[8] The *Faithfulness Condition* can be thought of as the assumption that conditional independence relations are due to the causal structure rather than to accidents in parameter values [20].

where D is the data, M are all the possible structures, and M_R are all the structures that contain the relation R. Thus, $P(M)$ denote the prior probability of a structure M and $P(D \mid M)$ the probability of the data given the structure. Calculating 4.15 is very costly, so it is approximated by the marginal likelihood of the data given the structure, and usually restricted to a maximum number of variables in the network.

Depending on the reliability threshold, the resulting network can have undirected edges, $-$, which means that there is not enough information to obtain the direction of the link, and bi-directed edges, $\leftrightarrow$, indicating that there is a common cofounder. So in general, BCCD returns a PAG.

4.3.4.3 FCI Algorithm

Fast Casual Inference (FCI) is an extension of the PC algorithm for causal structure learning, considering variables that are causally insufficient. The Fast Causal Inference Algorithm [20] tolerates, and sometimes discovers, unknown confounding variables, and it has been shown to be asymptotically correct even in the presence of confounders.

Before describing the FCI algorithm, we need to define some graphical concepts. Given a DAG over the set of variables $\mathbf{V}$, where $\mathbf{O} \subset \mathbf{V}$ are the observed variables:

Collider: B is a collider along the path $< A, B, C >$ if and only if $A* \rightarrow B \leftarrow *C$.

Definite non-collider: B is a definite non-collider on undirected path U if and only if either B is an endpoint of U, or there exist vertices A and C such that U contains one of the subpaths $A \leftarrow B * - * C$, $A * - * B \rightarrow C$, or $A * - * \underline{B} * - * C$. ($A * - * \underline{B} * - * C$ means that edges A, B and A, C do not collide in B.)

Edge into A: An edge between B and A is into A if and only if $A \leftarrow *B$.

Edge out of A: An edge between B and A is out of A if and only if $A \rightarrow B$.

Inducing path: an undirected path between $A, B, \in \mathbf{O}$ is an *inducing path* relative to $\mathbf{O}$ if and only if every member of $\mathbf{O}$ in the trajectory is a collider except the endpoints, and every collider is an ancestor of either A or B. It can be shown that if there is an inducing path between A and B, they can not be d-separated by any subset of variables of $\mathbf{O}$; and if there is not such inducing path, there is at least one subset of $\mathbf{O}$ that d-separates them [20].

Definite discriminating path: In a partially oriented inducing path graph, U is a definite discriminating path for B if and only if U is an undirected path between $X \neq B$ and $Y \neq B$ containing B, every vertex on U except for B and the endpoints is a collider or a definite non-collider on U, and (i) if V and V' are adjacent on U, and V is between V and B on U, then $V* \rightarrow V$ on U; (ii) if V is between X and B on U and V is a collider on U then $V \rightarrow Y$, else $V \leftarrow *Y$; (iii) if V is between Y and B on U and V is a collider on U then $V \rightarrow X$, else $V \leftarrow *X$; and (iv) X and Y are not adjacent.

Triangle: A triangle in a graph G is a complete subgraph of G with three vertices; i.e., vertices X, Y, Z form a triangle if and only if X and Y are adjacent, Y and Z are adjacent and X and Z are adjacent.

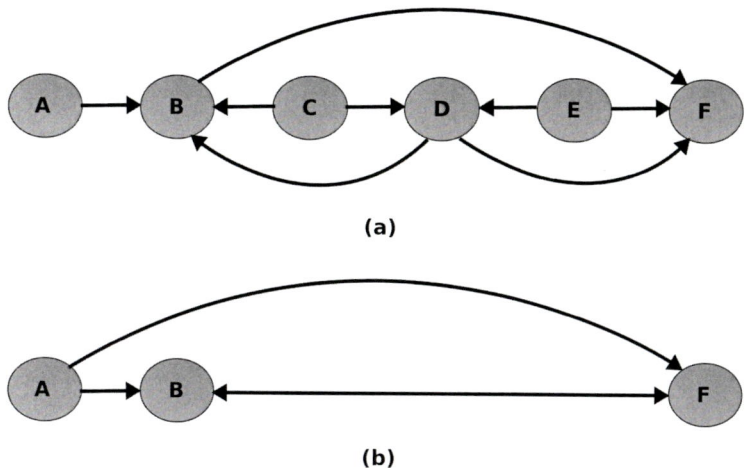

Fig. 4.9 Given the DAG in (a), the path $< A, B, C, D, E, F >$ is an inducing path over $O = \{A, B, F\}$; in (b) it is shown the corresponding induced path graph over O. Figure based on [20]

Inducing path graph: G' is an *inducing path graph* over **O** for a DAG G if and only if **O** is a subset of the vertices in G, there is an edge between variables A and B with an arrowhead at A if and only if A and B are in **O**, and there is an inducing path in G between A and B relative to **O** that is into A. In an inducing path graph, there are two kinds of edges: $A \rightarrow B$ implies that every inducing path over **O** between A and B is out of A and into B, and $A \leftrightarrow B$ entails that there is an inducing path over **O** that is into A and into B. This last kind of edge can only occur when there is a latent common cause of A and B [20]. An example of an inducing path and inducing path graph is depicted in Fig. 4.9.

In what follows, we consider the same types of marks as in PAGs for the edges: arrowhead ($>$), tail ($-$) or circle ($\circ$); plus an ($*$) that indicates any of the previous marks. Sepset(X, Y) contains the subsets of variables that d-separate X, Y, and Adjacencies(G, X) the variables that are adjacent to X in the graph G. To estimate d-separation from the probability distributions, the FCI algorithm performs conditional independence tests for discrete variables or partial correlations for linear, continuos variables. Next, we present the Causal Inference (CI) algorithm (antecedent of FCI):

1. Build a complete undirected graph G over the set of variables V.
2. If $A, B \in V$ are d-separated given any subset S of V, remove the edge between A and B, and record S in Sepset(A, B) and Sepset(B, A).
3. Let G' be the graph resulting from step (2). Orient each edge as $\circ - \circ$. For each triplet of vertices A, B, C such that the pair A, B and the pair B, C are each adjacent in G' but the pair A, C are not adjacent in G', orient $A * - * B * - * C$ as $A* \rightarrow B \leftarrow *C$ if and only if B is not in Sepset(A, C), and orient it as

$A * - * \underline{B} * - * C$ if and only if B is in Sepset(A, C). ($A * - * \underline{B} * - * C$ means that edges A, B and A, C do not collide in B.)

4. Repeat:

- If there is a directed path from A to B, and an edge $A * - * B$, orient $A * - * B$ as $A* \rightarrow B$,
- Else if B is a collider along the path $< A, B, C >$ in G', B is adjacent to D, and D is in Sepset(A, C), then orient $B * - * D$ as $B \leftarrow *D$,
- Else if U is a definite discriminating path between A and B for M in G', and P and R are adjacent to M on U, and $P - M - R$ is a triangle, then:

  - If M is in Sepset(A, B) then M is marked as a noncollider on subpath $P * - * M * - * R$,
  - Else $P * - * M * - * R$ is oriented as $P* \rightarrow M \leftarrow *R$.

- Else if $P* \rightarrow M * - * R$ then orient as $P* \rightarrow M \rightarrow R$.

until no more edges can be oriented.

The CI algorithm is not practical for large numbers of variables because of the way the adjacencies are constructed. While it is correct to remove an edge between A and B from the complete graph if and only if A and B are d-separated given some subset of V, this is impractical for two reasons: (i) there are too many subsets of V on which to test the conditional independence of A and B, (ii) for discrete distributions, unless the sample sizes are enormous, there are no reliable tests of independence of two variables conditional on a large set of other variables.

The FCI algorithm follows a strategy that is similar to PC to avoid testing independence on all subsets of V. Given the initial complete undirected graph, it removes an edge between X and Y if they are d-separated given subsets of vertices adjacent to X or Y in G. This will eliminate many, but perhaps not all of the edges that are not in the inducing path graph. Second, it orients edges by determining whether they collide or not, just as in the PC algorithm. In a third step, FCI finds possible additional d-separations not found in the first step.

The FCI algorithm:

1. Build a complete undirected graph G over the set of variables V.
2. $n = 0$
   Repeat:

   - Repeat:

     - Select an ordered pair of variables X and Y that are adjacent in G such that Adjacencies(G, X) has a cardinality greater than or equal to n, and a subset S

of Adjacencies(G, Y) of cardinality n, and if X and Y are d-separated given S delete the edge between X and Y from G, and record S in Sepset(X, Y) and Sepset(Y, X).

until all ordered variable pairs of adjacent variables X and Y such that Adjacencies(G, X) has a cardinality greater than or equal to n and all subsets S of Adjacencies(G, X) of cardinality n have been tested for d-separation;
- $n = n + 1$

until for each ordered pair of adjacent vertices X, Y, Adjacencies(G, X) has a cardinality smaller than n.
3. Let G' be the undirected graph resulting from step (2), orient each edge as $\circ - \circ$. For each triplet of vertices A, B, C such that the pair A, B and the pair B, C are each adjacent in G', but the pair A, C are not adjacent in G', orient $A *-* B *-* C$ as $A* \rightarrow B \leftarrow *C$ if and only if B is not in Sepset(A, C).
4. For each pair of variables A and B adjacent in G', if A and B are d-separated given any subset S of Possible-D-SEP(A, B) or any subset S of Possible-D-SEP(B, A) in G remove the edge between A and B, and record S in Sepset(A, B) and Sepset(B, A).

FCI then reorients an edge between any pair of variables X and Y as $X \circ - \circ Y$, and proceeds to reorient the edges in the same way as steps (3) and (4) of the CI algorithm.

An example of a model (a PAG) obtained with FCI algorithm is shown in Fig. 4.10, considering that the independence tests are correct. For more details and the theoretical bases of the CI and FCI algorithms see [20].

4.3.5 Casual Discovery with Functional and Parametric Constraints

In general it is not possible to obtain a unique causal model, so the previous methods return a CPDAG or a PAG. However, if some additional assumptions are made about the functional and/or parametric forms of the underlying causal structure, we can exploit asymmetries in order to identify the direction of a structural relationship. These asymmetries manifest in various ways, including non-independent errors, measures of complexity, and dependencies between marginals and cumulative distribution functions.

A case where this asymmetries are present is the one of linear models with the additive noise. If the noise is non-Gaussian, and we want to determine the direction of the arc $X - Y$, the distribution of the residuals when estimating the regression of one variable onto the other, allows us to determine the direction of the arc. For instance, if the causal direction is $X \rightarrow Y$, then residuals of the regression of Y onto X are uncorrelated with X, but the residuals of the regression of X onto Y are

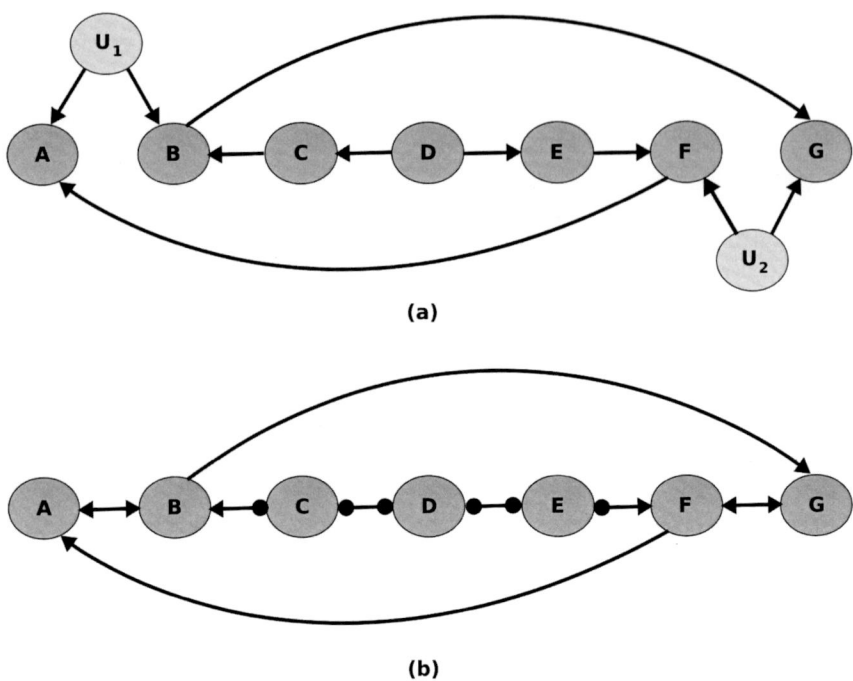

Fig. 4.10 **a** Original causal Bayesian network, where A, B, C, D, E, F, G are the observed variables and U_1, U_2 are unobserved or hidden variables. **b** The PAG that will be obtained by the FCI algorithm assuming correct independence tests. Figure based on [20]

dependent on Y. LiNGAM, an algorithm that takes advantage of this asymmetry, is described below.

4.3.5.1 LiNGAM

LiNGAM (Linear, Non-Gaussian, Acyclic Model) [18] proposes a different approach for causal discovery considering linear non-Gaussian models. It is based on a statistical tool to identify linear models: *Independent Component Analysis* (ICA) [5]. In a linear model, the variables $\mathbf{X} = \{X_1, ..., X_N\}$ are linear functions of their parents and the error terms, here represented by e, which are assumed to be continuous random variables with non-Gaussian distributions and independent of each other.

LiNGAM assumes: (i) causal sufficiency, (ii) linear relationships, (iii) non-Gaussian noise with non-zero variance. It also considers a causal ordering of the variables, $X_1, X_2, ...X_N$, such that a variable X_i can not be a cause of a variable X_j, $j < i$ according to the causal order; which is not required to be known *a priori* and can be estimated . The structural equation of each variable is of the following form:

$$X_i = \sum_j b_{ij} X_j + e_i + c_i \qquad (4.16)$$

where b_{ij} are constants, e_i is a non-Gaussian noise term and c_i is an optional constant term. The noise terms, e_i, are independent with non-zero variance.

In the LiNGAM algorithm the causal model is represented by a matrix equation:

$$\mathbf{X} = \mathbf{B}\mathbf{X} + e \qquad (4.17)$$

where the matrix $\mathbf{B}$ is an $N \times N$ matrix representing the edge coefficients on the corresponding graph. If the columns of $\mathbf{B}$ correspond to the hierarchical order of the variables in the graph, then $\mathbf{B}$ is lower triangular. The basic idea of LiNGAM is to recover the $\mathbf{B}$-matrix from a data matrix. Solving the above equation for $\mathbf{X}$ we get:

$$\mathbf{X} - \mathbf{B}\mathbf{X} = e$$

So:

$$\mathbf{X} = (\mathbf{I} - \mathbf{B})^{-1}e \qquad (4.18)$$

Which can be written as:

$$\mathbf{X} = \mathbf{A}e \qquad (4.19)$$

where $A = (\mathbf{I} - \mathbf{B})^{-1}$.

Equation 4.19 together with the above assumptions fits the ICA framework, and it follows that $\mathbf{A}$ is identifiable. ICA is able to estimate $\mathbf{A}$, however with indetermination regarding the permutations of the variables and a scale factor. ICA returns $W_{ICA} = \mathbf{PDW}$, where $\mathbf{P}$ is an unknown permutation matrix and $\mathbf{D}$ a diagonal matrix.

To determine the correct permutation, $\mathbf{P}$ should be so that $\mathbf{DW}$ should have no zeros in the main diagonal; given that $\mathbf{B}$ should be a lower triangular matrix with zeros in the main diagonal and $\mathbf{W} = \mathbf{I} - \mathbf{B}$. The correct scale can be determined by using the ones in the diagonal of $\mathbf{W}$. To obtain $\mathbf{W}$ it is only necessary to divide the rows of $\mathbf{DW}$ by the corresponding diagonal elements. Finally the weight matrix, $\mathbf{B}$, is obtained as $\mathbf{B} = \mathbf{I} - \mathbf{W}$. The non-zero terms in $\mathbf{B}$ determine the structure of the causal graphical model.

LiNGAM algorithm:

1. Given a d-dimensional random vector $\mathbf{X}$ and a matrix of observed data of dimension $d \times n$, apply an ICA algorithm to obtain an estimation of the matrix $\mathbf{A}$.
2. Find a permutation of the rows of $\mathbf{W} = \mathbf{A}^{-1}$ to generate a matrix $\mathbf{W}'$ without zeros in the main diagonal.
3. Divide each row of $\mathbf{W}'$ by the corresponding element in the main diagonal to obtain a new matrix $\mathbf{W}''$ in which all the elements in the main diagonal are 1.
4. Obtain an estimate $\mathbf{B}'$ of $\mathbf{B}$ such that $\mathbf{B}' = \mathbf{I} - \mathbf{W}''$.
5. Finally, to estimate the causal order, find the permutation matrix $\mathbf{P}'$ of $\mathbf{B}'$, such that the matrix $\mathbf{B}'' = \mathbf{P}'\mathbf{B}'\mathbf{P}'^T$ is as close as possible to a lower triangular matrix.

Thus, the assumptions of linearity and non-Gaussian independent error terms, enables the discovery of the exact true causal structure, as opposed to only the Markov equivalence class.

4.3.6 Continuous Optimization

Previous techniques for causal discovery rely on combinatorial optimization. Recently, there have been an increasing number of methods which seek to learn structure from data, while leveraging the advantages of continuous optimization. Continuous optimization methods are common in the field of deep learning, where highly parameterized networks are optimized using variations of gradient descent. Increased computational power, particularly with the use of GPUs, makes the task of learning from large, high-dimensional datasets feasible.

Continuous optimization approaches reformulate the combinatoric graph-search problem into a continuous optimization problem. In general, these methods seek the adjacency matrix G^* that minimizes some score function $S(G)$, as in the combinatorial approaches, but this time subject to the constraint $h(A) = 0$. Here, h is the function used to enforce that inferred graph is acyclic.

A review of several continuous optimization approaches is included in [22], below we briefly describe some representative examples.

4.3.6.1 DAGs with NO TEARS

DAGs with NO TEARS [24] is considered as the first to recast the combinatoric graph search problem as a continuous optimization problem. They propose the following function for enforcing acyclicity:

$$h(\mathbf{A}) = tr(e^{\mathbf{A} \odot \mathbf{A}}) - d = 0 \tag{4.20}$$

where tr is a threshold, $\mathbf{A}$ is the adjacency matrix of dimension d.

A disadvantage of this acyclicity constraint is that the matrix exponential requires $O(d^3)$ computations, and subsequent methods seek to improve on this. The structural model learnt is linear, although noise variables are not assumed Gaussian. It uses least-squares loss with an $l1$ penalty to encourage sparsity, and their objective is optimized using the Augmented Lagrangian method. Despite the fact that the formulated optimization problem does not guarantee an optimal solution, their results demonstrate close-to-optimal solutions.

4.3.6.2 Causal Generative Neural Network

Causal Generative Neural Network (CGNN) [7] combines graph learning with continuous optimization, neural networks, and hill-climbing search. The neural networks are used to learn the functions mapping variables, where the variables themselves are selected according to the output of a greedy-search algorithm. The networks are trained using the Adam optimizer with a Maximum Mean Discrepancy (MMD) score function. During training, the edges are directed in order to minimize the discrepancy; following training, the graph is adjusted to remove cycles. CGNN incorporates a hill-climbing search algorithm to optimize the structure of the DAG, and then the network optimization resumes. This training cycle is repeated until convergence, and each edge has an associated score representing its contribution to the global fit. They use a thresholding function to regularize the number of edges in the graph.

There are other continuous optimization methods that can handle non-parametric models, for learning time-series, and for causal discovering with latent variables, see [22].

4.4 Learning Subject-Specific Causal Models

In some domains there are variations on causal relations across the subjects of a population. In neuroscience, for example, it has been observed that causal relations between the brain regions could vary across patients because of differences in their recovery or the degree of disease affectation. Investigations in genetics have found that there are within an individual tumor, specific causal relations between somatic genome alterations and differential expressed genes. Methods for learning subject-specific causal models could be beneficial for these domains, allowing them to make better decisions. A challenge in learning-subject specific models, is that usually there is very limited data for each subject. A way to approach this challenge is to use *transfer learning*, which has shown to be useful for learning models with limited data sets, allowing the use of related data sets with different different, although similar, probability distributions.

KTL-WeFGES [17] is a knowledge transfer method for learning subject-specific causal models from limited observational data sets. It is a scored-based method that alleviating the lack of data in the target domain, includes strategies for using additional data from related domains, weighted according to their similarities in probability distribution with target subject data.

The problem is defined as follows: Given a target domain D_T and a set of S source domains $\{D_s\}$, $s = 1, .., S$, how to improve a target causal model (MEC), combining the data from D_T and $\{D_s\}$, where $\forall_s : V_T = V_s$. Where V_T is the set of variables in the target domain and V_s the set of variables in each source domain.

The method uses weighted instances of the source data sets for evaluating the local structure of candidate target MECs. These weights represent the relevance or similarity between the source and target domains. In the case of data with a multinomial distribution, the method gives higher weights to those source instances that are more similar in conditional probability distribution to the target. The similarity for each local probability distribution is based on the Kullback-Leibler divergence (KLD) that estimates the difference between $P_T(x \mid Pa(x))$ and $P_s(x \mid Pa(x))$, for all variables x in the target and source data sets, where $Pa(x)$ are the parents of x in the causal models (an initial causal model is obtained for the target domain, which later is improved with the transferred data). The KLD is obtained by summing overall the variables:

$$KLD(D_T, D_s) = \sum_x P_T(x \mid Pa(x)) log \frac{P_T(x \mid Pa(x))}{P_s(x \mid Pa(x))} \qquad (4.21)$$

Then the weight for all instance in $\{D_s\}$ is calculated as:

$$W_s(D_T, D_s) = exp[\frac{-(1/rq)KLD(D_T, D_s)}{\alpha_w}] \qquad (4.22)$$

where $1/rq$ is a normalization factor over the number of possible configurations for X and $Pa(X)$, which avoids increments in the KLD when the number of parents for X increases; and $\alpha_w = (1/S)\sum_s KLD(D_T, D_s)$, which normalizes the weights.

For linear Gaussian models, the similarity is estimated based on the probability distribution *discrepancy* (see [17]).

Causal discovery is based on the FGES algorithm, with the difference that the total score of a potential MEC is obtained by summing over its scores from the target and sources data sets combination, where the sources data is weighted according to the scheme previously described. Experiments with synthetic and benchmark causal Bayesian networks show a significant improvement when transferring data from related domains according to several metrics [17]. The difference with respect to learning only from the target data is higher when the target sample size is smaller, and they tend to converge when the data in the target domain increases. Thus, this is useful for learning subject-sepecific models with very limited data.

4.5   Determining Undefined Edges

In general, causal discovery algorithms obtain a MEC, so some edges remain undirected. SLICE (Structural Learning with Intervals of Causal Effects) [12], is an algorithm to decide on undirected edges, based on the computation of causal effects. It is based on the hypothesis that casual directions could be inferred from the assessment of the strength of potential causal effects and such assessment can be computed by intervals comparison strategies

SLICE assumes linear models and causal sufficiency. In a linear model when the causal structure is known, for example, $X \rightarrow Y$, the causal effect of X on Y can be obtained by the regression coefficient, β, of X in Y. In this case $\beta \neq 0$ (in general, $\beta \neq 0$ if the observed variable is not in the parents set of the intervened variable). For the reverse effect, when Y is intervened and X is observed, the causal effect is zero. It follows that the causal effect computed in the true causal direction should result in a greater causal effect than the one computed in the reverse of the true causal direction.

For the case in which the unique causal structure is not known (we have a CPDAG), it is possible to compute causal effects but, instead of obtaining a scalar, a multi-set of potential causal effects is computed. To compute the causal effects in this case where the DAG is not known, the IDA (Intervention calculus when the DAG is absent) algorithm is used [9], which returns a multi-set of causal effects. Multi-sets can be characterized in terms of their lower and upper bounds. SLICE considers these bounds as *intervals of causal effects* (ICE) for their comparison.

Let $X - Y$ be a CPDAG representing an equivalence class of an unknown causal structure, and let ICE_1 and ICE_2 be two intervals of causal effects computed by manipulating X and observing Y and vice versa, obtained from the potential structures $X \rightarrow Y$ and $X \leftarrow Y$, respectively. The true causal direction between X and Y can be found by identifying whether $ICE_1 > ICE_2$ or $ICE_1 < ICE_2$.

Intervals are compared based on an *acceptability index*. Given two intervals, A and B, with lower limits A_L and B_L and upper limits A_R and B_R, pre-ordered by their midpoints, $m(A) \leq m(B)$, the acceptability of $A < B$ is:

$$Acc(A < B) = \frac{m(B) - m(A)}{w(A) + w(B)} \tag{4.23}$$

where $m(A) = (a_L + a_R)/2$, $m(B) = (b_L + b_R)/2$; and $w(A) = (a_R - a_L)/2$, $w(B) = (b_R - b_L)/2$ are the half-widths of the corresponding intervals.

The acceptability index can be classified in three cases according to the following expression:

$$Acc(A < B) \begin{cases} = 0 & m(A) = m(B) \\ > 0 \wedge < 1 & m(A) < m(B) \wedge a_R > b_L \\ \geq 1 & m(A) < m(B) \wedge a_R \geq b_L \end{cases} \tag{4.24}$$

If $Acc(A < B) = 0$, we reject that A is inferior to B; if $0 < Acc(A < B) < 1$, A is inferior to B to some extent with uncertainty; and if $Acc(A < B) \geq 1$ we have absolute certainty that B is superior to A.

SLICE assumes that an equivalence class of causal structures, **G**, has been obtained by any causal discovery algorithms. The following procedure summarizes SLICE:

1. Extract the set, T, of undefined edges, $X - Y$, in **G**.
2. For each edge $\in T$, calculate the multi-set of causal effects of intervening X over Y and viceversa.
3. For each multi-set, obtain the upper and lower bounds, and based on these calculate the acceptability index which are stored in Acc_{set}.
4. Select the undefined edge T_0 that has the maxima in Acc_{set}.
5. Orient the edge $X - Y$ in T_0 according to the direction given by the interval comparison (greater causal effect).
6. Apply Meek rules to orient other edges, if possible.
7. Update **G** and T.
8. Repeat 2–7 until no edge remains undirected.

Experiments with synthetic CBNs, show that SLICE can recover the actual correct causal graph with relatively small errors (according to the structural Hamming distance), in particular for linear models with Gaussian noise. In the case of linear models with non-Gaussian noise, LiNGAM tends to get better results.

4.6 Additional Reading

A comprehensive discussion of causal graphical models, and in particular of the
PC and FCI algorithms, can be found in [20]. Peters et al. [14] focuses on the
analysis of cause-effect relations for two variables, as well as its relation to machine
learning. Causal strcuture learning for linear non-Gaussian models is described in
[18]. The BCCD algorithm for learning causal graphs is introduced in [3]. A review
of alternative approaches for learning causal networks can be found in [6], another
review with focus on continuous optimization is [22].

4.7 Exercises

1. Given the following conditional independence relations: $I(Y, X, Z)$, $I(X,$
   $YZ, W)$ and $\neg I(Y, W, Z)$: (a) draw the CPDAG that represents these relations,
   (b) draw all the DAGs in the Markov equivalence class.
2. Given the graph in Fig. 4.11, draw all the DAGs that are in the same Markov
   equivalence class.
3. Different DAGs can be associated to a MAG. A particular one is called a *canonical
   DAG*, D, and is obtained from a MAG by the following transformations: (i) $X \rightarrow$
   Y in the MAG, then $X \rightarrow Y$ in D; (ii) $X \leftrightarrow Y$ in the MAG, then $X \leftarrow U_{XY} \rightarrow Y$
   in D. Show the canonical DAG for the MAG in Fig. 4.6b.
4. Consider the BN structure of the golf example given in Fig. 4.11 and the associated
   data (Table 4.2), calculate the BDeU score for this example.
5. Given the same data as in the previous problem, and an alternative structure for the
   golf example that is in the same Markov equivalence class, calculate the BDeU
   score. Is it the same as in the previous problem?

Fig. 4.11 A BN structure for
the golf example

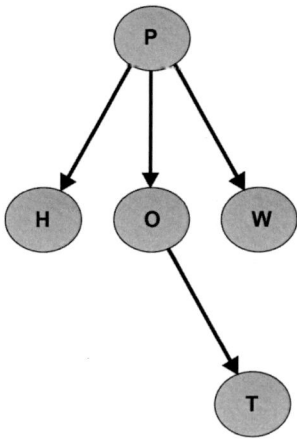

Table 4.2 Data for the golf example

Outlook (O)	Temperature (T)	Humidity (H)	Wind (W)	Play (P)
Sunny	Medium	High	No	N
Sunny	High	High	No	N
Overcast	High	High	No	P
Rainy	Medium	High	No	P
Rainy	Low	Normal	Yes	P
Rainy	Low	Normal	Yes	N
Overcast	Low	Normal	Yes	P
Sunny	Medium	High	No	N
Sunny	Medium	Normal	No	P
Rainy	Medium	Normal	No	P
Sunny	Medium	Normal	Yes	P
Overcast	Medium	High	Yes	P
Overcast	High	Normal	Yes	P
Rainy	Medium	High	Yes	N

6. Given the data on the golf example, apply the standard PC algorithm to learn the structure of a BN. Then, apply the BCCD algorithm, using an estimate of the reliability of each causal link to order the causal relation, and setting a threshold to determine if a link has certain direction or remains undirected. Compare the structures obtained with PC and BCCD.

7. Given the DAG in Fig. 4.9a, obtain the inducing paths over $O = \{A, B, D, E, F\}$ and $O = \{A, B, D, F\}$.

8. Consider a linear system with 4 variables $X_1, ...X_4$, and non-Gaussian noise $e_1, ...e_4$. Given the following equations: $X_1 = X_4 + e_1$, $X_2 = 0.2X_4 + e_2$, $X_3 = 2X_2 + 5X_1 + e_3$, $X_4 = e_4$. (a) Write the equations in matrix form, $\mathbf{X} = \mathbf{BX} + e$, reordering the variables so the matrix $\mathbf{B}$ is in lower triangular form. (b) Draw the corresponding graphical model.

9. *** Develop a program that implements the PC algorithm for discrete variables considering the variation that makes it independent of the variable order.

10. *** Obtain data from some real-world domain and use the program from the previous exercise to obtain a causal model. Compare the structures obtained with the two variantes (original, order independent) of the PC algorithm.

Acknowledgements Section 4.2 and some parts of Sect. 4.3 are based on [21], used with permission from Springer. The section on continuous optimization is a summary of [22].

References

1. Alonso-Barba JI, Gámez JA, Puerta JM et al (2013) Scaling up the greedy equivalence search algorithm by constraining the search space of equivalence classes. Int J Approximate Reasoning 54(4):429–451
2. Chickering DM (2002) Optimal structure identification with greedy search. J Mach Learn Res 3:507–554
3. Claassen T, Heskes T (2012) A bayesian approach to constraint based causal inference. In: Proceedings of Uncertainty in Artificial Intelligence (UAI). AUAI Press, pp 207–216
4. Colombo D, Maathuis MH (2014) Order-independent constraint-based causal structure learning. J Mach Learn Res 15:3921–3962
5. Comon P (1994) Independent component analysis—a new concept? Signal Process 36:287–314
6. Daly R, Shen Q, Aitken JS (2011) Learning bayesian networks: approaches and issues. Knowl Eng Rev 26(2):99–157
7. Goudet O, Kalainathan D, Caillou P, Guyon I, Lopez-Paz D, Sebag M (2018) Learning functional causal models with generative neural networks. arXiv:1709.05321v3
8. Heckerman D, Geiger D, Chickering DM (1995) Learning Bayesian networks: the combination of knowledge and statistical data. Mach Learn 20(3):197–243
9. Maathuis MH, Kalisch M, Buhlmann P (2009) Estimating high-dimentional intervention effects from observational data. Ann Stat 37(6A):3133–3164
10. Meek C (1995) Causal inference and causal explanation with background knowledge. In: Proceedings of the eleventh conference on Uncertainty in artificial intelligence, pp 403–410
11. Messerli FH (2012) Chocolate consumption, cognitive function, and nobel laureates. New England J Med 367(16):1562–1564
12. Montero-Hernández S, Orihuela-Espina F, Sucar LE (2018) Intervals of causal effects for learning causal graphical models. Proc Mach Learn Res 72:296–307
13. Ogarrio JM, Spirtes P, Ramsey J (2016) A hybrid causal search algorithm for latent variable models. In: Conference on probabilistic graphical models, pp 368–379
14. Peters J, Janzing D, Scholkpf B (2017) Elements of causal inference. MIT Press
15. Ramsey J, Glymour M, Sanchez-Romero R, Glymour C (2017) A million variables and more: the fast greedy equivalence search algorithm for learning high-dimensional graphical causal models with an application to functional magnetic resonance images. Int J Data Sci Anal 3(2):121–129
16. Reichenbach H (1956) The direction of time. University of California Press, Berkeley, CA
17. Rodríguez-López V, Sucar LE (2022) Knowledge transfer for causal discovery. Int J Approximate Reasoning 143·1–25
18. Shimizu S, Hyvärinen A, Kano Y, Hoyer PO (2005) Discovery of non-gaussian linear causal models using ICA. In: Proceedings of the 21st conference on uncertainty in artificial intelligence (UAI-05). AUAI Press, pp 525–533
19. Sokolova E, Groot P, Classen T, Heskes T (2014) Causal discovery from databases with discrete and continuous variables. In: Probabilistic graphical models (PGM). Springer, pp 442–457
20. Spirtes P, Glymour C, Scheines R (2001) Causation, prediction, and search, 2nd edn. MIT Press
21. Sucar LE (2021) Probabilistic graphical models: principles and applications. Springer Nature, Switzerland
22. Vowels MJ, Camgoz NC, Bowden R (2022) D'ya like DAGs? a survey on structure learning and causal discovery. ACM Comput Surv 55(4):1–36
23. Zhang K (2024) Causal representation learning: uncovering the hidden world. In: 2nd workshop on causal discovery. IBERAMIA
24. Zheng X, Aragam B, Ravikumar P, Xing EP (2018) DAGs with NO TEARS: continuous optimization for structure learning. arXiv:1803.01422v2

Causal Discovery from Interventional Data

5

Abstract

In this Chapter we present an overview of how to discover the causal structure through interventions. We start by providing a formal definition of an intervention and describing the different types of interventions. Then we formalize the definition of an experiment and the assumptions that the search methods usually make. Finally, we describe fixed and adaptive search strategies to recover the causal structures via experiments, and provide worst case bounds on the required number of experiments for the different types of interventions.

5.1 Introduction

Given a set of causal variables, we can distinguish two principal ways to learn about causal relations between these variables. We can either obtain passive observational measurements of the variables or we can subject a subset of the variables to an experimental intervention. In the first case, the values of the variables are recorded as they naturally occur. In the second case, we intervene on the natural process by controlling the values of some of the variables and record the values of the others.

© The Author(s), under exclusive license to Springer Nature Switzerland AG 2026
L. E. Sucar, *Causal Discovery*, Computer Science Foundations and Applied Logic,
https://doi.org/10.1007/978-3-031-98345-0_5

Fig. 5.1 Causal graphical
structures that represent the
conditional independence
relation $I(X, Y, Z)$

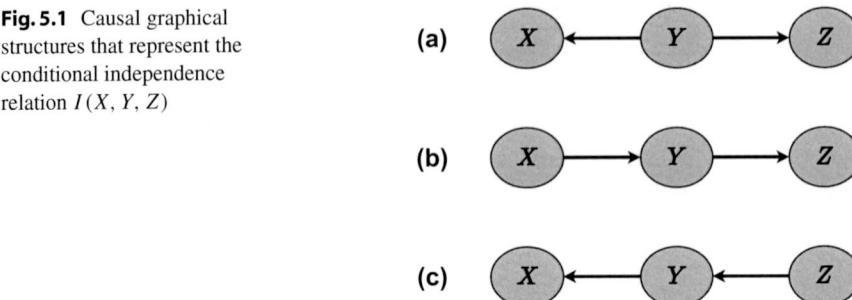

There are also hybrid methods, that combine observations with interventions. The
use of interventions in this way for causal discovery is found in randomized experiments in clinical trials, but in general the notion of interventions is much broader.
In the previous chapter we reviewed several algorithms for causal discovery from
observational data, in this chapter we will focus on how to learn a causal structure
from interventional data.

As we discussed in the previous chapter, causal discovery from observational data
is limited, in general, to obtaining the Markov equivalence class; interventions are
required to identify the true causal graph. Interventions provide, in addition to the
passive observational distribution over the variables, a manipulated distribution. Different graphs, that appear equivalent given observational data, can be distinguished
by their different manipulated distributions. For example, consider the three causal
structures depicted in Fig. 5.1; they all represent the same conditional independence
relation, $I(X, Y, Z)$, so they can not be distinguished from observational data (unless
additional assumptions are introduced). However, if we manipulate Y, we can distinguish between the three alternatives depending on the causal effect on the other
two variables: (i) structure (a) if the causal effect is on the other variables, X and Z,
(ii) structure (b) if the causal effect is only on Z, and (iii) structure (c) if the causal
effects is only on X.

To recover the causal structure several experiments have to be performed. This
sequence of experiments have to be designed so that the *true* causal structure can be
obtained, and the number of required experiments can be minimized; that is, aim not
only for accuracy but also for efficiency.

Next we describe the different types of interventions, and then the search strategies
to minimize the number of required experiments.

5.2 Interventions

In previous chapters we defined the *do* operator to differentiate interventions from
observations (see Section 2.5), and we studied how to reason about interventions in
casual Bayesian networks (see Section 3.1). In this Section, we will go deeper about
interventions, discussing different types of interventions from different perspectives.

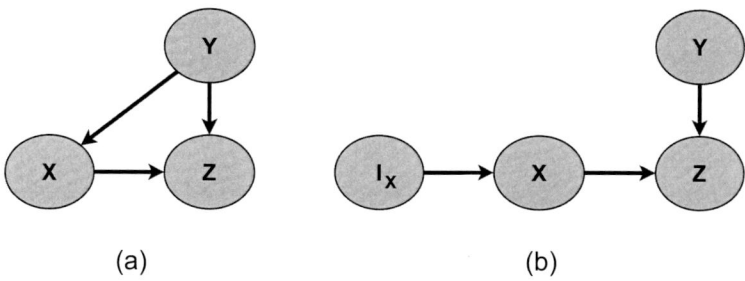

(a)   (b)

Fig. 5.2 **a** A causal graph. **b** The modified graph resulting from a structural intervention on X

Although there are different definitions of interventions (see [3]), here we will focus on those introduced by Pearl [7] and Spirtes [8].

Pearl considers interventions in relation to a structural equation model that describes the causal relations over a set of variables. An intervention is *atomic* if the intervention consists on lifting the intervened variable X_i from the functional equation $x_i = f(pa_i, u_i)$, and placing it under the influence of a new mechanism that sets the value of x_i, $do(X_i = x_i)$, while keeping all other mechanisms unperturbed. This can be represented graphically by an intervention variable that is added to the causal model pointing to X_i. Also, the intervention breaks the causal influence of other variables over the intervened variable, so the edges into the intervened variable are removed from the graph. The result of these two alterations to a graphical causal model are illustrated in Fig. 5.2. By specifying appropriate conditions on the interventions, such as clamping of the variable to a particular value, and the exogeneity of the intervention, the true causal structure can be discovered.

Fisher [4] developed the theory of randomized controlled trials for their value for causal discovery. He describes a randomization as an intervention that assigns treatment at random after all disturbing causes have been determined. The randomization is supposed to ensure that no influences, other than the intervention, determine the state of the intervened variable, so any confounding due to unknown common causes can be eliminated. This principle is one of the guiding considerations when using interventions as a discovery tool.

5.2.1 Types of Interventions

Eberhardt [3] distinguishes two types of interventions: (i) structural interventions and (ii) parametric interventions. Correa and Bareinboim [1] consider three types of interventions: (i) atomic, (ii) conditional, and (iii) stochastic. Atomic are the same as structural interventions, and stochastic correspond to parametric interventions. In conditional interventions, the intervened variables is replaced with a deterministic function of some observable variables. In this chapter we will focus on structural and parametric interventions.

5.2.1.1 Structural Interventions

Structural interventions make the intervened variable independent of its normal causes; this type of interventions are sometimes referred to as randomizations [4], surgical interventions [7], or ideal interventions [8]. Given a set of measured variables $\mathbf{V}$, a structural intervention I_s on a subset $\mathbf{S} \subseteq \mathbf{V}$ is an intervention on $\mathbf{S}$ that satisfies the following constraints [3]:

1. No variable in $\mathbf{V}$ is a cause of I_s.
2. I_s makes every variable in $\mathbf{S}$ independent of its causes, and it determines the distribution of $\mathbf{S}$; that is, the term $P(\mathbf{S} \mid pa(\mathbf{S}))$ is replaced by $P(\mathbf{S} \mid I_s)$, and all other terms are left unchanged in the factored joint probability distribution of $\mathbf{V}$.

The definition of a structural intervention implies that the causal structure is manipulated, since any causal influence on the intervened variable is destroyed. The manipulated causal structure is referred to as the *post-manipulation graph*. The post-manipulation graph is the causal graph where all the edges incident on any of the intervened variables are removed. An example of a structural intervention is shown in Fig. 5.2. These changes in the causal structure imply that also the joint probability distribution is changed, as we saw in previous chapters.

5.2.1.2 Parametric Interventions

Parametric interventions are a *weaker* form of intervention, also known as partial, soft, conditional or dependent intervention. Given a set of measured variables $\mathbf{V}$, a parametric intervention I_p on a subset $\mathbf{S} \subseteq \mathbf{V}$ is an intervention on $\mathbf{S}$ that satisfies the following constraints [3]:

1. No variable in $\mathbf{V}$ is a cause of I_p.
2. I_p does not make the variables in $\mathbf{S}$ independent of their causes (i.e., does not eliminate the edges incident on $\mathbf{S}$). Its effect is in the factored joint probability distribution, $P(\mathbf{V})$, where the term $P(\mathbf{S} \mid pa(\mathbf{S}))$ is replaced by $P(\mathbf{S} \mid pa(\mathbf{S}), I_p = k)$, and all other terms are left unchanged.

Although a parametric intervention does not imply any structural changes among the variables in the causal graph, and the post-manipulation graph is only changed by the addition of the intervention variables, its influence is evident in the manipulated probability distribution. Parametric interventions do not alter the structure, instead they modify the parameters of the conditional distribution of the intervened variables.

As an example, consider the classical causal model relating *smoking* and *cancer*, including a possible cofactor, *genotype*, see Fig. 5.3a. Consider that *smoking* represents the number of cigarets a person smokes per day. A parametric intervention could be to increase this number, let's say by 10. The resulting model is depicted in Fig. 5.3b. This will change the probability distribution of *cancer* given *smoking*.

It is not necessary for a parametric intervention to consist of adding a constant to the value of the intervened variable; all that is required is that there is a change in the

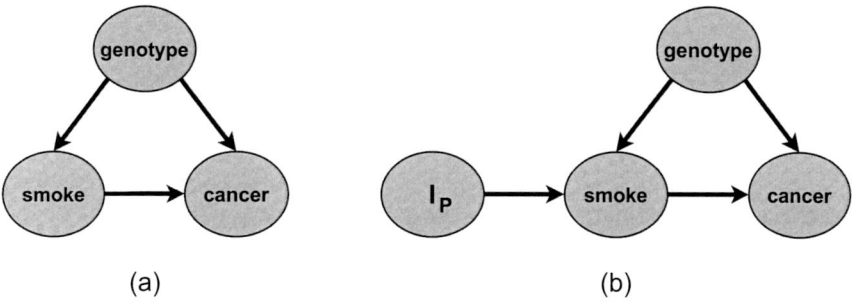

Fig. 5.3 a A causal model relating smoking and cancer. **b** The modified model after a parametric intervention on *smoke*

conditional probability distribution from the passive observational one. The difficulty of parametric interventions lies in determining how they should be performed when nothing is known about the causal structure among the variables.

5.2.1.3 Instrumental Variables

Parametric interventions are closely related to the idea of *instrumental variables*. Consider that there are two variables, X and Y, which are related by a linear model, $Y = \beta X + \epsilon$, where ϵ is an error term. If the error term affects both, X and Y (cofactor), then β can not be directly estimated. In such case, a consistent estimator of β can be found if there is a variable Z (called an instrument) that is correlated with X, independent of ϵ, and not directly correlated with Y.

This structure mirrors the set-up for parametric interventions, as shown in Fig. 5.3b, where I_p (corresponding to Z) could be an instrumental variable. The difference between the two is mainly in the semantics: instrumental variables are generally real variables, corresponding to some causally relevant feature in the real world, whereas intervention variables are just part of the model.

Structural and parametric interventions are the two extremes on a continuum of harder to weaker interventions. Structural interventions make the intervened variable independent of its causes, while parametric interventions only affect the parameterization of the causal model. There is also the possibility of mixed interventions. In this case, the manipulated distribution is a mixture of two manipulated distributions, one structurally manipulated and one parametrically manipulated.

The interventions we have discussed are all designed for static models; they do not work for dynamic (time series) models. In dynamic models we have to consider how fast, and for how long, the effect of interventions percolates through the system, and the frequency at which the system is being sampled. Also, we have to distinguish between a one-time intervention and a continuously occurring intervention.

5.3 Searching for a Causal Structure Through Interventions

Given the definitions of interventions, now we will focus on how to select the best next experiments to perform when searching for a causal structure.

How to select the sequence of experiments to perform is given by a *search strategy*. Within *pure* strategies, we can distinguish two types: fixed and adaptive. A fixed search strategy determines a particular sequence of experiments before any data is collected. No adjustments are allowed after that, the full sequence of experiments must be performed. An adaptive strategy also defines a sequence of experiments prior to the first experiment, but the specification of the sequence of experiments can change depending on the previous experiments' results. An adaptive strategy can specify in advance how it will adapt in light of the particular outcomes of the previous experiments.[1]

5.3.1 Structure Learning Based on Bayesian Search with Interventions

Previous work on learning Bayesian networks, consider a Bayesian approach to structure learning based information theoretic measures to identify the optimal next experiment [9]. Their approach assumes a prior distribution $P(G)$ over the space of directed acyclic graphs over a set of N variables, and a prior over the parameterization for each graph $P(\theta \mid G)$, both of which are updated given data D from an experiment E that manipulates a subset of the variables V in the graph. This means that, for each possible graph $g \in G$, the manipulated graph is computed given the intervention set of the experiment E. Then, for each graph G_E, the parameters that are not manipulated by the experiment are updated given the data, and lastly, with the updated parameters in place, the distribution over the graphs can be updated. The posterior distribution over graphs after one experiment becomes the prior distribution for the next experiment.

The next experiment is selected so that the entropy of the posterior distribution over the possible graphs is minimized. However, the computation cost required to update the distribution is prohibitive, as the number of possible structures (DAGs) grows super-exponentially with the number of variables. Thus, approximations are used in practice, such as assuming a total ordering of the variables, or using sampling techniques.

---

[1] A mixed strategy can specify for each possible scenario (history of experiments and current state) a distribution over the possible experiments. A random sample from this distribution then determines the next experiment in the sequence, see [3].

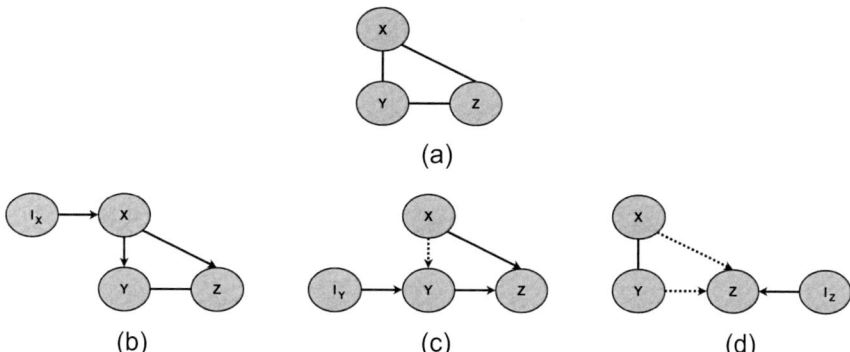

Fig. 5.4 An example of the results of an intervention in a causal model with three variables. **a** The skeleton that indicates a complete graph without the direction of the edges (assuming the correct structure has arcs directed as: $X \rightarrow Y$, $X \rightarrow Z$, and $Y \rightarrow Z$). **b** The result after an intervention on X. **c** Resulting graph after an intervention on Y. **d** The model resulting from an intervention on Z. We can observe that, in the three different interventions, we can not recover a unique causal model, there is at least one edge that remains undefined. Undirected edges mean that the two variables are adjacent but the orientation remains unknown. Dashed arrows mean that either there is an incoming edge or there is no edge at all

5.3.2 Search with Interventions: Assumptions

In general, it is not possible to determine a unique causal structure with a single intervention. For instance, consider a complete causal structure with three variables, as shown in Fig. 5.4. Figure 5.4a depicts the skeleton, and Figs. 5.4b, c, d show the resulting partial causal graphs after performing a structural intervention on each of the three variables. We can observe that, in the three cases, we can not obtain a unique and complete causal structure. In this example, a second intervention is sufficient to determine the causal structure; and, in general, a sequence of experiments is required.

Before we present the search strategies, we need to formalize the definition of an experiment and the assumptions that the search methods usually make (according to [3]).

Experiment: An experiment E_i on a set of variables **V** is represented by a triplet of sets $E_i = (\mathbf{S_i}, \mathbf{R_i}, \mathbf{I_i})$, where $\mathbf{S_i}$ represents the subset of variables in the model, **V**, that is subject to an intervention in E_i, $\mathbf{I_i}$ is the corresponding set of intervention variables, and $\mathbf{R_i}$ contains the remaining passively observed variables, such that $\mathbf{R_i} \cup \mathbf{S_i} = \mathbf{V}$.

It is possible that an experiment involves multiple independent, but simultaneous, interventions on a set of variables. We say that an experiment is a single intervention experiment if S_i is a singleton set, and a multiple intervention experiment if $\mathbf{S_i}$ contains several variables.

The search methods are sensitive to the assumptions regarding the causal models and the search space. The methods we will describe below, make the following assumptions [3]:

1. Causal Markov Condition—Every variable in **V** is independent of its nondescendents given its parents in the graph.
2. Causal Faithfulness—The probability distribution $P(\mathbf{V})$ is faithful to the graph G if all and only the independence relations true in $P(\mathbf{V})$ are entailed by the causal Markov condition.
3. Acyclic causal structure—The causal structure over a set of variables is acyclic.
4. Known distribution family—The model is a discrete binary model or a linear model with normal errors (it is known which).
   Other assumptions that are made by some methods are:
5. Causal sufficiency—There are no latent common causes of the set of variables **V**.
6. Oracle—The experiment returns the independence relations true in the manipulated population distribution.
7. Independence tests—Independence tests are the only admissible means to identify causal structure given a distribution over the variables.

A search strategy is a complete plan of which experiment to be performed next at any point in the sequence of experiments, and for any history of experimental outcomes. As mentioned before, we consider two classes of strategies, fixed and adaptive; and provide bounds on the number of experiments necessary to discover the causal structure.

5.3.3 Fixed Search Strategies

Fixed search strategies specify one particular sequence of experiments that do not depend on the outcomes of previous experiments. Fixed strategies can be used to identify the longest sequence of experiments that may be required, providing a worst-case bound for a search strategy. Next, we describe different strategies under different types of interventions and assumptions.

Structural Interventions
Considering all the above assumptions (1 to 7) and only a single structural intervention per experiment, we can derive the following bound on the necessary and sufficient number of experiments with N variables [2]: $N - 1$ experiments are sufficient, and in the worst case necessary, to determine the causal graph among $N > 2$ variables. For $N = 2$, two experiments are sufficient and in the worst case necessary.

A search strategy the guarantees the previous bound is the following.

Strategy 1: Fixed, Single Structural Intervention, Causally Sufficient—Given N variables, $X_1, ..., X_n$, the sequence of experiments, $E_1, ...E_{n-1}$, consists on subjecting each variable $X_1, ..., X_{n-1}$ to a structural intervention I_i such that $S_i = X_i$.

This strategy is not sensitive to the order of the variables subject to intervention, nor does it matter which particular variable is not intervened (note that $N - 1$ variables are subject to an intervention).

For example, consider the example in Fig. 5.4. If we initially intervene X we obtain the partial causal graph in Fig. 5.4b. A second intervention, in either Y or Z will determine the direction of the undirected edge. Thus, two interventions are sufficient to determine the causal structure in this example with three variables.

The above results apply when we assume causal sufficiency; however, for a causally insufficient set of variables, no sequence of experiments is sufficient to determine the causal graph among N variables when only a single structural intervention is permitted in each experiment.

If we apply multiple interventions we can reduce the number of required experiments.

Strategy 2: Fixed, Multiple Structural Interventions, Causally Sufficient—Given N variables, $X_1, ..., X_n$, the sequence of experiments, $E_1, ...E_k$, is such that for each pair of variables X, Y one of the following holds:

1. There is at least one experiment where X is subject to an intervention and Y is not, and one where Y is subject to an intervention and X is not.
2. There is at least one experiment where X is subject to an intervention and Y is not, and one where both X and Y are passively observed.

For a causally sufficient system, $\lfloor log_2 N \rfloor + 1$ experiments are sufficient, and in the worst case necessary, to determine the causal graph among N variables when multiple simultaneous and independent structural interventions are performed in each experiment.

There are several ways to make interventions that satisfy the previous constraints. For instance, one option is that in each experiment $\lfloor N/2 \rfloor$ variables are subject to an intervention, with different combinations of variables in each experiment.

When we perform multiple interventions then it is possible to obtain the structure considering causally insufficient systems.

Strategy 3: Fixed, Multiple Structural Interventions, Causally Insufficient—Given N variables, $X_1, ..., X_n$, the sequence of experiments, $E_1, ...E_N$, is such that in each experiment a simultaneous structural intervention is performed on $N - 1$ variables, leaving out a different variable in each experiment.

For a causally insufficient system, N experiments are sufficient, and in the worst case necessary, to determine the causal graph among the N observed variables when multiple simultaneous and independent structural interventions are performed in each experiment. Note that only the structure between the observed variables is recovered.

It is possible to discover the causal graph with other strategies that do not require a simultaneous intervention on $N - 1$ variables for each experiment, but for such strategies more than N experiments are required, and at least one of them must still intervene all but one variable.

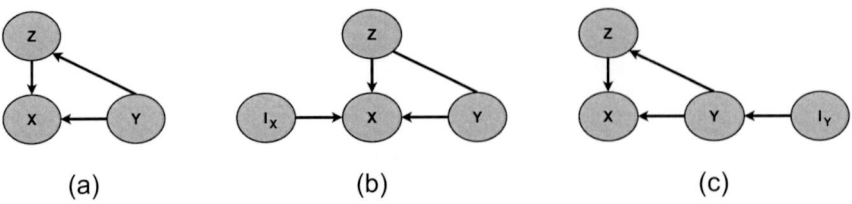

Fig. 5.5 An example of a series of experiments with parametric interventions. **a** The correct causal model. **b** The resulting graph after the first experiment, a parametric intervention on X. **c** The resulting causal graph after the second experiment, a parametric intervention on Y

Parametric Interventions
Parametric interventions modify the parameters of the causal model, and do not imply any structural changes among the variables; thus, they can be used more efficiently than structural interventions, resulting in different bounds on the number of required experiments.

In the case of single parametric interventions, there is no difference with structural interventions (Strategy 1), just applying parametric instead of structural interventions.

Strategy 4:   Fixed, Single Parametric Intervention, Causally Sufficient—Given N variables, $X_1, ..., X_n$, the sequence of experiments, $E_1, ...E_{n-1}$, consists on subjecting each variable $X_1, ..., X_{n-1}$ to a parametric intervention I_i such that $S_i = X_i$.

Considering all the above assumptions (1 to 7) and only a single parametric intervention per experiment, $N - 1$ experiments are sufficient, and in the worst case necessary, to determine the causal graph among N variables.

For instance, consider a causal graph with three variables, Fig. 5.5a shows the true causal model. If we perform a parametric intervention on X, Fig. 5.5b, we can test for independence $I(I_X, X, Z)$, and we find that they are not independent, then we can orient the edge from Z towards Y (a collider). Similarly for $I(I_X, X, Y)$, so we orient the edge from Y towards X. With this experiment we can not determine the edge $Z - Y$, so we need a second experiment. If we intervene Y, Fig. 5.5c, we can test for $I(I_Y, Y, Z)$, and we will find they are independent. As we know that $I_y \rightarrow Y$, then it must be that $Y \rightarrow Z$.

Strategy 5:   Fixed, Multiple Parametric Interventions, Causally Sufficient—Given N variables, $X_1, ..., X_n$, the sequence of experiments consists of one experiment E_1 in which all but one variable, $X_1, ..., X_{n-1}$, are subject to a parametric intervention.

By considering multiple simultaneous interventions it is possible to reduce the number of experiments with respect to structural interventions. One experiment is necessary and sufficient to determine the causal graphical structure for N variables when multiple simultaneous parametric interventions are performed in each experiment.

The great reduction in the number of experiments results from the fact that parametric interventions can be combined independently of each other, as they do not destroy the causal structure. However, a reduction in the number of experiments comes at a price: substantially more conditional independence tests may be required. In the single experiment of Strategy 5 all variables may have to be tested for colliders on the basis of the data, and the collider tests generally involve higher order independence tests, which are less reliable.

The previous results consider causal sufficiency, if we take out this assumption it becomes imposible to recover a single causal model with parametric interventions. That is, for causally insufficient systems, no sequence of experiments is sufficient to determine the causal graph among N variables if only parametric interventions (single or multiple) are allowed. However, in certain classes of structures, it is possible to recover the causal model, depending on the existence of certain *inducing paths*.[2]

Strategy 6:   Fixed, Single Parametric Intervention, Causally Insufficient—Given N variables, $X_1, ..., X_n$, the sequence of experiments, $E_1, ... E_n$, consists on subjecting each variable $X_1, ..., X_n$ to a parametric intervention I_i such that $S_i = X_i$.

In the case of causally insufficient systems and parametric interventions, we can recover the causal graph according to the following conditions. Let G be a causal graph over a set of variables $\mathbf{V}$, and let $\mathbf{O}$ be a subset of $\mathbf{V}$ containing the observed variables. Let G_m be a graph where each variable $X \in \mathbf{O}$ is extended with an intervention variable $I_X \to X$. The causal subgraph over the observed variables can be uniquely determined by parametric interventions on each variable in $\mathbf{O}$ if and only if for each pair of variables $X, Y \in \mathbf{O}$ that are non-adjacent, there are no inducing paths between I_X and Y and no inducing paths between I_Y and X in G_m.

Table 5.1 provides a summary of the required number of experiments, in the worst case, to recover a unique causal structure for fixed strategies with structural or parametric interventions.

5.3.4   Incorporating Structural Knowledge

If additional structural information is provided *a priori*, the number of required interventions can be reduced. This information may include, for a particular edge: (i) adjacency (connected but do no know the direction), (ii) non-adjacency, (iii) directed cause, (iv) semi-directed cause (non-adjacency or direct cause). For example, we might learn from observations a CPDAG (includes some adjacency and directed causes), and can perform some interventions to find the direction of the undefined

---

[2] See Sect. 4.3.4.3 for a definition of an inducing path.

Table 5.1 Number of experiments required in the worst case for fixed strategies. *Impossible* means that in general it is not possible to discover the causal structure, but there could be especial cases in which it is possible

Type of experiment	Causal sufficiency	No. of experiments (worst case)
Structural interventions		
Single	Yes	$N - 1$
Multiple	Yes	$log_2(N) + 1$
Single	No	Impossible
Multiple	No	N
Parametric interventions		
Single	Yes	$N - 1$
Multiple	Yes	1
Single	No	Impossible
Multiple	No	Impossible

Fig. 5.6 An example of a knowledge graph. Edges XY and YW are adjacent, XW a directed cause, and edge XZ is a semi-directed cause

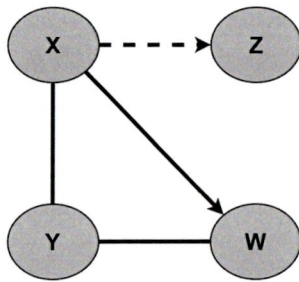

edges. A causal graph that contains this different types of information is known as a *knowledge graph* (KG). Figure 5.6 shows an example of a knowledge graph.

OPTINTER [3] is an algorithm that finds an intervention set given a knowledge graph. By calling this algorithm before each new experiment, the shortest sequence of experiments necessary given the initial knowledge graph is obtained. In the case that the knowledge graph is a CPDAG, if multiple simultaneous and independent structural interventions can be performed in each experiment, then $\lfloor log_2(|CL_{max}|) \rfloor$ experiments are sufficient to recover the true causal graph, where CL_{max} is the largest clique of undirected edges in the CPDAG [5]. This is an important result with practical implications, as we can obtain a CPDAG based only on observational data, and then perform a "few" interventions to recover a unique DAG.

The algorithm is based on the idea of orienting edges in the largest clique of undefined edges in the knowledge graph. A clique makes edge-orientations maximally independent, because fewer orientations are implied. Thus, to minimize the number of experiments, it must break down cliques of connected variables as fast as possible.

Given a set of **V** vertices and *MaxInt* the maximum size of the intervention set **S** for the next experiment, the OPTINTER algorithm is the following:

Fig. 5.7 An example of the OPINTER algorithm. **a** Initial knowledge graph. **b** Knowledge graph after an intervention on Z

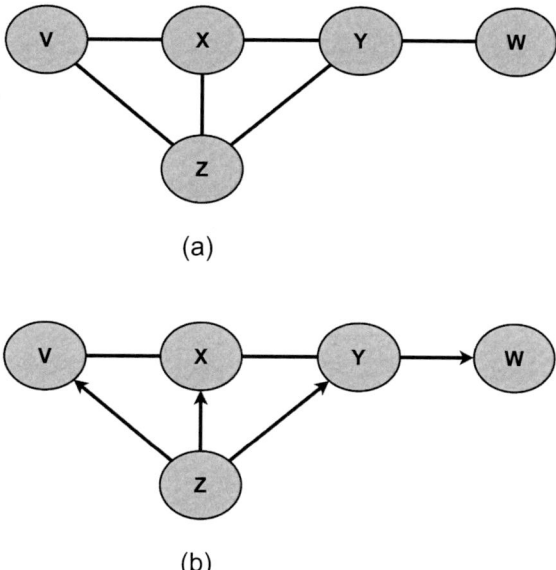

(a)

(b)

1. Initialization: make all vertices *admissible* and initialize the intervention set **S** to empty.
2. Find all the cliques of vertices connected by unknown edges (adjacent or semi-directed cause), and order them by the number of vertices, C_i, they have, mark them as *unresolved*.
3. The relevant cliques are those with $C_i > h$, where $h = 2^{\lceil log_2(|C_{max}|) \rceil - 1}$
4. Sort all relevant cliques in order of size, being C_{curr} the largest unresolved clique.
5. For each vertex $V_i \in C_{curr}$, set nc_i to the number of unresolved relevant cliques in C_i such that $V_i \in C_i$.
6. While ($|S| < MaxInt$) $\wedge$ ($|C_{curr} \cap S| < |C_{curr} - h|$), select vertex $V \in C_{curr}$ such that V is admissible and has the highest nc. Put it in **S**
7. Return to 4 and repeat until all relevant cliques are resolved or when no further relevant cliques can be resolved.
8. Return the intervention set **S**.

As an example, consider the KG (skeleton) in Fig. 5.7a. It has three cliques: $C_1 = \{V, X, Z\}$, $C_2 = \{X, Y, Z\}$, $C_3 = \{Y, W\}$. The largest cliques are C_1 and C_2. Lets say we select C_1, we then need to find the vertices in C_1 which are in more relevant cliques, in this case X and Z. We select one of this at random to include in the intervention set, let it be Z (in this example we are assuming interventions on a single variable in each experiment, $MaxInt = 1$). We then perform the experiment with the intervention on Z, and according to the independence constraints we obtain the KG in Fig. 5.7b. The orientation of the edge YW was implied by the orientation of the edge ZY resulting from the intervention. In this new graph, two cliques remain $C_1 = \{V, X\}$ and $C_2 = \{X, Y\}$. As they are of the same size we can choose any, so we select C_1; X is the only variable that is in both cliques, so we intervine X and according to the results of the experiment we will recover the unique causal graph.

5.3.5   Adaptive Search Strategies

The previous strategies belong to the category of fixed strategies; that is, they consider a predefined series of experiments independently from the results of previous experiments. An alternative are the *adaptive strategies*, in which the next experiment to perform is based on the results of the previous experiments. Although adaptive strategies can not reduce the number of experiments required in the worst case; in some cases, given the information from the previous experiments, it becomes possible to select the next experiment optimally, and therefore reduce the number of required experiments in contrast with the worst case scenario.

These strategies are particularly useful when additional information is obtained when performing an experiment; for example, independence relations obtained in one experiment that imply the orientation of some edges. In some cases an adaptive search strategy might reduce the required number of experiments, but in the worst case it is the same as the fixed search strategies.

One adaptive strategy is to start with a fixed search strategy, and after the first experiment, apply the OPTINTER algorithm:

Strategy 7:   Adaptive, Structural Interventions, Causally Sufficient—Given N variables, $X_1, ..., X_n$, in the first experiment, E_1, apply Strategy 1 or Strategy 2. For the subsequent experiments, $E_2, ...E_k$, apply the intervention set **S** given by OPINTER, according to the current KG obtained from the last experiment.

For more details regarding adaptive strategies see [3].

5.4   Additional Readings

The following thesis provides a thorough analysis on learning causal models from interventions [3]. The main accounts of interventions are given by Pearl [7], Spirtes et al. [8], and Woodward [10]. Given a MEC, [6] describes a framework for active learning of causal structures via intervention experiments. Other type of interventions, known as conditional interventions, are addressed in [1]; that also generalizes the do-calculus for general interventions. A comprehensive review on causal discovery is in [11], including observational and interventional data.

5.5 Exercises

1. Given the hypothetical causal model for *stroke* in Fig. 5.8, show the modified graph if we perform a simultaneous structural intervention in the variables *Lack of Exercise* and *Unhealthy diet*.
2. Given the hypothetical causal model for *stroke* in Fig. 5.8, and considering that the variable *Lack of Exercise* represents the number of hours of exercise per week, show the modified graph if we perform a parametric intervention on this variable.
3. Given the example in Fig. 5.4, determine the complete causal model by performing two experiments for the three cases, the first one is that in Fig. 5.4b, c, d and in the second experiment, perform an intervention on: (b) Y, (c) Z, (d) X. According to the result of this intervention, define the undefined edges.
4. Repeat the three cases of the previous problem, but with an intervention on the other variable. Verify that the results are consistent.
5. Given the hypothetical causal model for *stroke* in Fig. 5.8: (a) How many multiple structural interventions are necessary to determine the causal structure (assuming we do not know it)? (b) Specify a possible sequence of experiments.
6. Specify a set of variables to be intervened in each experiment, for a sequence of multiple structural interventions, considering: (a) $N = 7$, and (b) $N = 8$.
7. Given the example in Fig. 5.4a, consider that there is a hidden common cause between each pair of variables. (a) How many multiple structural interventions are required to obtain the causal structure between the observed variables? (b) Propose a sequence of experiments.
8. Consider the causal graph in Fig. 5.5a, and that we do not know the directions of the edges. Describe how the causal structure can be recovered by performing

Fig. 5.8 Hypothetical causal model for stroke

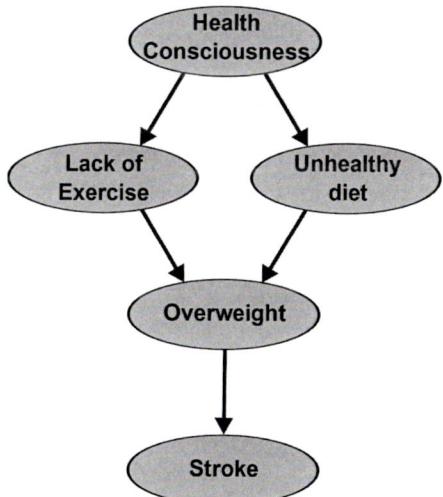

parametric interventions and independence tests (as in the example): (a) the first intervention is on Y, (b) the first intervention is on Z.

9. *** Investigate adaptive search strategies. Show some examples in which the required number of experiments can be reduced with respect to the worst case scenario.

10. *** Develop a program to test the search strategies on linear Gaussian causal models, by simulating the estimation of causal effects given structural interventions.

Acknowledgements This chapter is based on [3].

References

1. Correa JD, Bareinboim E (2020) A calculus for stochastic interventions: causal effect identification and surrogate experiments. In: Proceeding of the thirty-fourth AAAI conference on artificial intelligence

2. Eberhardt F, Glymour C, Scheines R (2006) N-1 experiments suffice to determine the causal relations among N variables. In: Holmes DE, Jain LC (eds) Innovations in machine learning, volume 194 of theory and applications series: studies in fuzziness and soft computing. Springer

3. Eberhardt F (2007) Causation and intervention, PhD Thesis, Carnegie Mellon University

4. Fisher R (1935) The design of experiments. Hafner

5. Hauser A, Bühlmann P (2012) Two optimal strategies for active learning of causal models from interventions. In: Sixth European workshop on probabilistic graphical models. Granada, Spain

6. He Y, Geng Z (2008) Active learning of causal networks with intervention experiments and optimal designs. J Mach Learn Res 9:2523–2547

7. Pearl J (2009) Causality: models, reasoning and inference. Cambridge University Press, New York

8. Spirtes P, Glymour C, Scheines R (2000) Causation, prediction, and search. MIT Press

9. Tong S, Koller D (2001) Active learning for structure in Bayesian networks. In: Nebel B (ed) Proceedings of the 17th international joint conference on artificial intelligence (IJCAI-01). Morgan Kaufmann, pp 863–869

10. Woodward J (2003) Making things happen. Oxford University Press (2003)

11. Zanga A, Ozkirimli E, Stella F (2022) A survey on causal discovery: theory and practice. Int J Approximate Reasoning 151:101–129

Causal Discovery from Temporal Data

6

Abstract

This chapter presents causal discovery from temporal data. It starts by introducing dynamic Bayesian networks, as a way to represent a causal dynamic structure. The main focus is on multivariate time-series, describing several techniques for causal discovering, including Granger causality, constraint-based and score-based approaches, and linear models with non-Gaussian noise. Then it analyzes causal discovery from event sequences. It finalizes by introducing the problem of subsampling, and different strategies to cope with this problem.

6.1 Introduction

Causal discovery from time series aims at discovering, from observational data, causal relations on temporal data, which can be categorized in two types: (i) multivariate time-series, and (ii) event sequences. Multivariate time-series consist of an N-variate time series $\mathbf{X}$ where, for a certain time t, each $\mathbf{X}_t$ is a vector $\mathbf{X}_t = (X_t^1, X_t^2, ..., X_t^N)$ in which each variable X_t^i represents a measurement of the $i - th$ time series at time t. Event sequences consist of multiple events, $\mathbf{E_t}$, which

© The Author(s), under exclusive license to Springer Nature Switzerland AG 2026   119
L. E. Sucar, *Causal Discovery*, Computer Science Foundations and Applied Logic,
https://doi.org/10.1007/978-3-031-98345-0_6

are organized in a specific order based on their time of occurrence. Causal discovery from temporal data enables better predictions and decisions based on a deeper understanding of a phenomena.

Inferring causal relations from time series data have served as the basis for causal discovery in various fields such as climate systems, ecological networks, effective connectivity in the brain, and finance [10, 13, 15]. One of the advantages of using observational data from time series is that the temporal order of the information can simplify causal analysis. That is, the causal driver can be identified as the variable that occurred first, as the future can not affect the past. This is known as the *Temporal Priority* [3]: *A causal relation between two variables is said to satisfy the temporal priority if it is oriented in such a way that the cause occurred before its effect.*

6.1.1 Temporal Data Configurations

Temporal data, depending on its intrinsic characteristics and data acquisition modalities, can be in different configurations:

- Type of data: multivariate time-series or event sequences.
- Variable values: can be discrete or continuous. Continuous data can assume any value in certain range. Discrete data can be further subdivided in ordinal (discrete values with an specific order) or nominal (no order).
- Time intervals: can also be discrete or continuous. Continuous time intervals mean that the data can be recorded at any point in time. Discrete time intervals refer to data recorded at regular, equally spaced time intervals.

In this chapter we will focus on temporal data obtained in discrete time intervals, with discrete or continuous values.

Next, we describe a graphical representation of time series based on dynamic Bayesian networks. Then, we present different approaches for causal discovery from multivariate time-series, followed by an introduction to causal discovery from event sequences, and lastly we discuss the problem of subsampling and how it can be solved.

6.2 Representation

To represent a dynamical causal structure obtained from multivariate, discrete intervals, time series data, we can use a graphical model known as a *Dynamic Bayesian Network* (DBN), which is an extension of Bayesian networks for dynamic domains. A DBN consists of a series of *time slices* that represent the state of all the variables at a certain time, t; a kind of snapshot of the evolving temporal process. For each temporal slice, a dependency structure between the variables at that time is defined, called the *base network*. It is usually assumed that this structure is duplicated for all

Fig. 6.1 An example of a dynamic Bayesian network with 3 variables and 4 time slices. In this case, the base structure is $X \to S \to E$, which is repeated across the 4 time slices

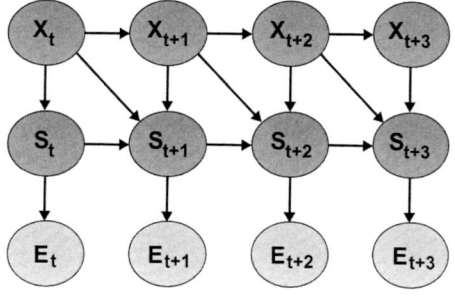

the temporal slices (except the first slice, which can be different). Additionally, there are edges between variables from different slices, with their directions following the direction of time, defining the *transition network*. Usually DBNs are restricted to have directed links between consecutive temporal slices, known as a first order Markov model; although, in general, this is not necessary. An example of a DBN with 3 variables and 4 time slices is depicted in Fig. 6.1.

Most of the DBNs considered in practice satisfy the following conditions:

- First order Markov model. The state variables at time t depend only on the state variables at time $t - 1$ (and other variables at time t).
- Stationary process. The structure and parameters of the model do not change over time.

There are three ways to represent graphically a DBN: (i) full-time causal graph, (ii) window causal graph, and (iii) summary causal graph or rolled graph. A full-time causal graph represents the model for the complete set of time intervals, although if this is very large, a subset is drawn. A window causal graph depicts a window that represent the necessary time intervals according to the order of the temporal model. If it is a first-order Markov model, two time slices are sufficient. An alternative for a more compact representation of a DBN is a summary causal graph or *rolled graph*, in which each variable is represented only once as a node in the graph, and an arc is drawn between the nodes that have direct causal relations. In the case of a model in which there could be links with different time lags, a number is associated to each arc indicating the corresponding time lag (when all the relations have a time lag of 1, this information can be omitted). The last two types of representations assume a stationary process. An example of the three types of representations is shown in Fig. 6.2.

6.3   Causal Discovery from Multivariate Time Series

Next we describe different types of algorithms for causal discovery from multivariate time series:

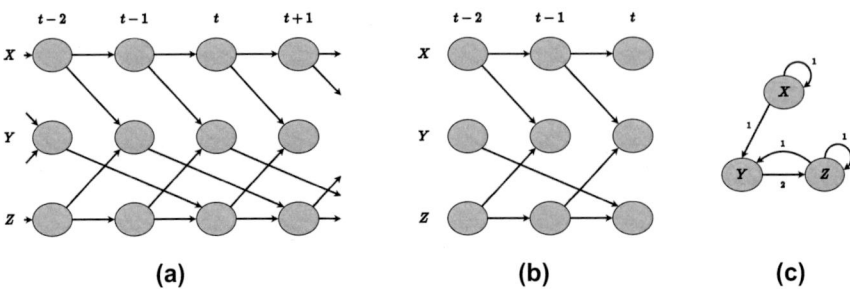

Fig. 6.2 An example of a dynamic Bayesian network represented as a: **a** full-time causal graph, **b** window causal graph, **c** rolled graph

- *Granger Causality*,
- Constraint-based algorithms,
- Score-based algorithms,
- Algorithms based on parametric/functional restrictions.

6.3.1 Granger Causality

Granger [10] proposed a practical definition of causality based on prediction improvement. The underlying idea of measuring whether *X causes Y* is that there is some information in *X* relevant for *Y* that is not contained in *Y*'s past as well as the past of "all the information in the universe" [10]. In practice, typically only *Y*'s past up to certain time is used. Granger causality is based on two principles:

1. The cause happens before its effect.
2. The cause has unique information about the future values of its effect.

An example is illustrated in Fig. 6.3, where we observe that some characteristic patterns in time series *X* are repeated after some time in *Y*.

Measuring prediction improvement can be operationalized in different ways. The most common framework are vector autoregressive models (VAR):

$$\mathbf{X}_t = \sum_{\tau=1}^{\tau_{max}} \Phi(\tau)\mathbf{X}_{t-\tau} + \eta_t \qquad (6.1)$$

where $\mathbf{X}_t = (X_t^1, X_t^2, ..., X_t^N)$ is a time series that contains N variables, $\Phi(\tau)$ is the $N \times N$ coefficient matrix at lag τ, τ_{max} some maximum time lag, and η denotes an independent noise term. Here, X_i causes X_j if any of the coefficients at lag τ is non-zero; which represents a causal link $X_i \rightarrow X_j$ at lag τ.

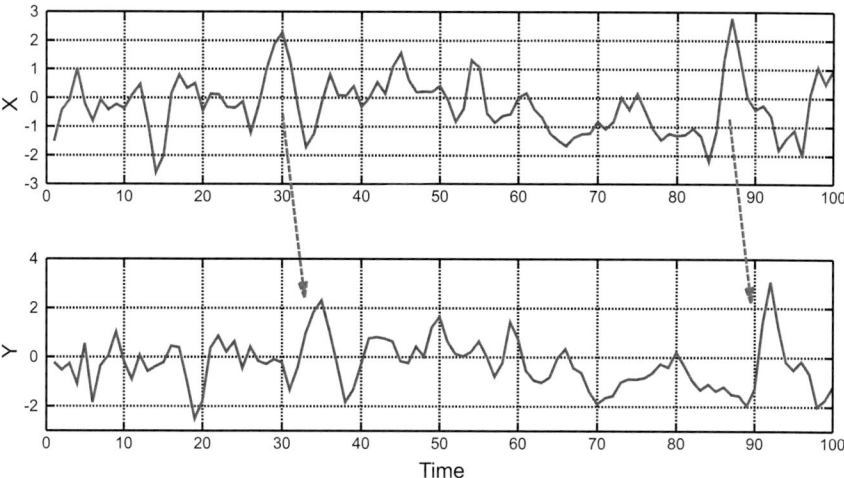

Fig. 6.3 An example of Granger causality. The variable Y seems to *follow* variable X, as some patterns in X repeat some time after in Y. So according to Granger causality, X *causes* Y

A more general definition is that of *transfer entropy*, which is based on the conditional mutual information (CMI), $I(X; Y \mid Z)$. Thus, tests on causality are based on testing whether a particular CMI is greater than zero.

$$I(X; Y \mid Z) = \sum_X \sum_Y \sum_Z P(x, y, z) log \frac{P(x, y \mid z)}{p(x \mid z) p(y \mid z)} \qquad (6.2)$$

If this test is different from zero for certain time lag τ, it implies a causal link $X_i \rightarrow X_j$ at lag τ; where Z represents Y's past and all other variables considered in the model.

Granger causality is not necessarily true causality. For instance, if both X and Y are driven by a common third process Z, with different lags, Granger causality might still consider that X causes Y. Yet, manipulation of X would not change Y. Granger-causality tests are designed to handle pairs of variables, and may produce misleading results when the true relationship involves three or more variables.

6.3.1.1   Multivariate Granger Causality

To overcome the problem of common confounders, all relevant information needs to be included. The multivariate Granger causality, or conditional Granger causality [9], makes use of the following full model, based on a vector autoregressive extension of the autoregressive model of the pairwise case:

$$X_t^q = a_{q,0} + \sum_{r=1}^{d} \sum_{i=1}^{\tau_{max}} a_{r,i} X_{t-i}^r + \eta_t^q \qquad (6.3)$$

where $\mathbf{X} = X_1, X_2, ...X_d$ is a d-dimensional time series, $a_{i,j}$ are real coefficients, τ_{max} is the maximum tal lag, and η_t^q are uncorrelated random variables with zero mean (noise terms). This version is sound and usually yields better results; however, its computational cost is such that in practice many studies rely on the pairwise version.

Other limitations of Granger causality are:

- It cannot deal with non-stationary processes.
- Its underlying linear assumption, as associations are highly likely to be non-linear on real datasets.
- It does not capture instantaneous causal relationships.

Several extensions have been proposed to deal with non-stationary processes and for non-linear systems, see [4,12,23].

6.3.2 Constraint-Based Approaches

Similarly to causal discovery for non-temporal data, constraint-based approaches exploit conditional independencies to build a skeleton between variables. This skeleton is then oriented according to a set of rules that define constraints on admissible orientations.

6.3.2.1 PCMCI Algorithm

The PCMCI (PC with Momentary Conditional Independence) algorithm is focused on causal discovery in time series [20]. It is an extension of the PC algorithm and it can detect lagged causal relations. It assumes a structural causal model with (non)linear relations and independent noise terms, as well as a stationary process. PCMCI relaxes the first-order Markov assumption and can detect causal relations at different time lags, although a maximum time lag needs to be specified.

PCMCI solves some limitations of the PC algorithm, in particular the processing time in data sets with high dimensionality, through the selection of conditions to eliminate irrelevant variables. This is achieved through the implementation of three stages:

1. A partially connected graph is built, connecting al pairs of nodes, X_{t-i}^p, X_t^q such that $X_{t-i}^p \to X_t^q$; $0 < i < \tau_{max}$ (τ_{max} refers to the maximum time lag considered for causal relations).
2. The second step removes all unnecessary edges based on conditional independence tests, as done in the PC algorithm, and takes into account the assumption of consistency throughout time to remove homologous edges. This uses a simplified (and faster) variant of the skeleton phase of the PC algorithm, called PC1, to learn a conditioning set that contains the parents for all variables X_i.

Fig. 6.4 An example of the PCMCI algorithm with two variables and three temporal stages. The graph illustrates the initial model that consists of a complete graph with a maximum time lag of two (1st stage)

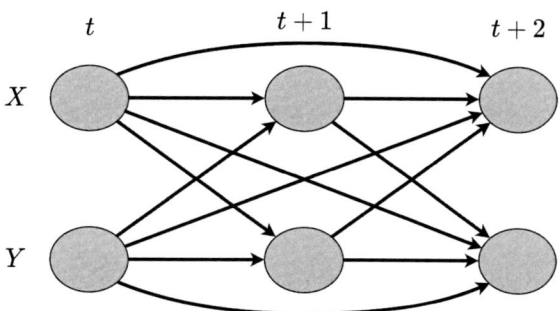

3. As the conditioning is based only on the parents of X_t^q, it cannot control for false positives with large autocorrelations in X_{t-i}^p . The third step deals with these autocorrelations by using the Momentary Conditional Independence test (MCI). MCI conditions on the parents of X_t^q and the parents of X_{t-i}^p while testing $X_{t-i}^p \rightarrow X_t^q$.

The MCI is used as an estimator for causal strength, as it quantifies the causal effect on X_t^q of an hypothetical perturbation in $X_{t-\tau}^p$. Thus, the value of the MCI statistics allows to rank causal links. The computational time is polynomial in the number, N, of time series and the maximum lag, τ_{max}.

As an example, consider two time series, X and Y, and a maximum time lag $\tau_{max} = 2$. In the first stage of the PCMCI algorithm, it generates a graph that includes all causal links $X_{t-i} \rightarrow Y_t$ and $Y_{t-i} \rightarrow X_t$, $i = 1, 2$. The resulting causal graph is depicted in Fig. 6.4. In the second stage, it applies conditional independence tests (as in PC) staring from a conditioning set, S, with $|S| = 0$ (marginally independent), and the increasing $|S|$ such that $|S| = |S| + 1$, until a maximum size. After this stage, it obtains the graph in Fig. 6.5. Finally, it applies the MCI test (third stage), and eliminates the dashed arcs in Fig. 6.5, obtaining as the final causal model the graph with only the continuous arcs. The corresponding SCM is:

$$X_t = f(X_{t-1}, Y_{t-1})$$
$$Y_t = f(X_{t-2}, Y_{t-1})$$
(6.4)

Instantaneous causal relations, which were not supported in the initial algorithm, have been integrated in a posterior version [21] by conducting separately the edge removal for lagged conditioning sets and instantaneous conditioning sets. Lagged relations are treated as in the original PCMCI algorithm and instantaneous relations are inferred using the PC algorithm.

6.3.2.2   LPCMCI Algorithm

The PCMCI algorithm assumes causal sufficiency. LPCMCI (Latent PCMCI) [8] extends PCMCI to take into account latent variables. In LPCMCI, known parents are used as default conditions whereas non-ancestors are not tested in conditioning

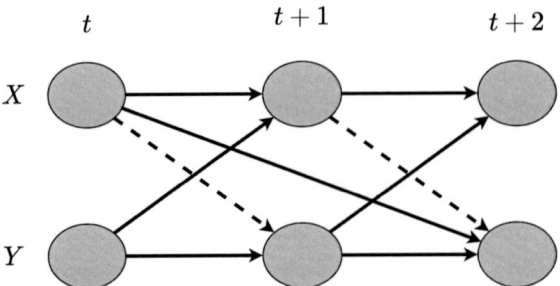

Fig. 6.5 An example of the PCMCI algorithm with two variables and three temporal stages. The graph illustrates the graphical model after applying the conditional independence tests (2nd stage). The dashed arrows are eliminated by the MCI test (third stage), so only the continuous arcs remain

sets. Furthermore, a new type of edge, with a middle mark, is used to facilitate early orientation of edges. In a preliminary phase, ancestors are detected during the skeleton construction through additional orientation rules adapted from those of the FCI algorithm (see Sect. 4.3.4.3). Then, in a final phase, edges are reoriented using the same rules. This algorithm is order independent, sound and complete.

6.3.3   Score-Based Approaches

In score-based approaches, a causal structure is searched among the possible struc-tures such that is has the *best-match* with the data, where best-match is based on a score that typically strikes a balance between the likelihood of the data given the network and a penalty term related to the complexity of the network. Compared to constraint-based approaches, score-based approaches have the advantage of assign-ing a score to the network inferred, a score that can then be used to assess the validity of the network. However, the solution obtained is in general suboptimal as finding a globally optimal network is known to be NP-hard. In addition, hidden variables have to be *postulated* (proposed by the causal discovery method, and their parameters can be obtained, for instance, by the Expectation-Maximization algorithm), and are not discovered as in constraint-based methods.

In principle we can apply methods for learning dynamic Bayesian networks, so we first present an overview of this type of methods; and then describe some technique developed specially for causal dynamic graphical models.

6.3.3.1   Learning Dynamic Bayesian Networks

Learning dynamic Bayesian networks involves two aspects: (i) learning the structure or graph topology, and (ii) learning the parameters or CPTs for each variable. Addi-tionally, we can consider two cases in terms of the observability of the variables: (a) full observability, when there is data for all the variables, and (b) partial observability,

Table 6.1 Learning dynamic Bayesian networks: 4 basic cases

Structure	Observability	Method
Known	Full	Maximum likelihood estimation
Known	Partial	Expectation–Maximization (EM)
Unknown	Full	Search
Unknown	Partial	EM and search

Fig. 6.6 Learning a DBN: first we obtain the base structure (left), and then the transition structure (right)

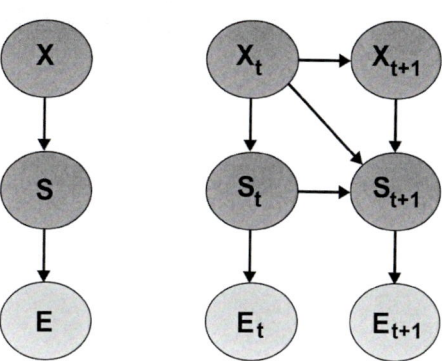

when some variables are unobserved or hidden. There are 4 basic cases for learning DBNs, see Table 6.1.

For the case of Unknown structure and Full observability, assuming that the DBN is stationary (time invariant), we can consider that the model is defined by two structures: (i) the base structure, and (ii) the transition structure. Thus, we can divide the learning of a DBN into two parts, first learn the base structure, and then, given the base structure, learn the transition structure, see Fig. 6.6.

For learning the base structure we can use all the available data for each variable, ignoring the temporal information. For learning the transition network we consider the temporal information, in particular the data for all variables in two consecutive time slices, X_t and X_{t+1}. Considering the base structure, we can then learn the dependencies between the variables at time t and $t + 1$ (assuming a first–order Markov model), and restricting the direction of the edges from the past to the future.

An algorithm for learning dynamic Bayesian networks that considers hidden variables is based on Structural Expectation-Maximization (SEM) [7]. This method considers stationary and Markovian (first-order) time series and uses a Bayesian score –Bayesian information criterion (BIC). At each iteration, the overall process consists in first improving the parameters of the base and transition networks, and then searching over DBN structures using expected counts to select the best scoring structures. The search over DBN structures is based on heuristics that typically consider neighboring structures of a given structure, obtained by arc additions, removals and reversals.

6.3.3.2 Dynotears

Pamfil et al. [18] proposed a method, named DYNOTEARS, which simultaneously estimates instantaneous and time-lagged relationships between time series through the adjacency matrices $\mathbf{W}, \mathbf{A}_1, \ldots, \mathbf{A}_{\tau_{max}}$, that represent the importance of the relation between two time series at different time lags; $\mathbf{W}$ for time lag 0 (instantaneous relation), and $\mathbf{A}_1, \ldots, A_{\tau_{max}}$ from a time lag of 1 to a τ_{max}. These matrices are learned by minimizing the penalized loss based on the Frobenius norm of the residuals of a linear model:

$$f(W, A) = \frac{1}{2d(T + 1 - \tau_{max})} ||X_t - X_t^T \mathbf{W} - X_{t+1:t+\tau_{max}} \mathbf{A}||_F^2 + \lambda_{\mathbf{W}} ||\mathbf{W}||_1 + \lambda_A ||\mathbf{A}||_1 \quad (6.5)$$

where ^{T} denotes transpose, $||.||$ represents the element-wise l_1 norm, and λ_W and λ_A are regularization constants.

The causal graph is then obtained by successively considering all relations at different time lags. To avoid cycles, an acyclicity constraint on the instantaneous adjacency matrix W is used. First, the best sparse DAG is selected using a score. A thresholding step is then used to prune some spurious correlations.

6.3.3.3 Fask

Sanchez-Romero et al. [22] made use of a variant of the PC algorithm, known as the Fast Adjacency Search stable (FAS-stable), to build a skeleton in which pairwise rules are used to orient edges. The method is known as FASK, for Fast Adjacency Skewness. The FAS-stable algorithm is an order independent adjacency search that avoids spurious connections between parents of variables. FAS-stable builds an undirected graph by iteratively testing conditional independencies; the BIC criterion is used for this testing. The orientation of two adjacent nodes X_p and X_q in the graph obtained is then based on a score comparing the conditional correlation of X_p and X_q given $X_p > 0$: if $corr(X_p, X_q \mid X_p > 0) > corr(X_p, X_q \mid X_q > 0)$, then $X_p \rightarrow X_q$; otherwise, $X_q \rightarrow X_p$.

6.3.4 Algorithms Based on Functional and Parametric Restrictions

As with non-dynamic causal models, if we incorporate functional and/or parametric restrictions, we can recover a unique causal structure. Structural Equation Models describe a causal system by a set of equations, where each equation explains one variable of the system in terms of its direct causes and some additional noise. In the case of continuous-valued data with the assumptions that the structural equation model is linear, acyclic, with non-Gaussian error terms, and assuming no hidden confounders, the full structure can be identified from observational data (LiNGAM, see Sect. 4.3.5.1.) Next, we describe a temporal extension of LiNGAM.

6.3.4.1 VarLiNGAM

VarLiNGAM [14] is a temporal extension of LiNGAM based on a structural vector autoregressive model where the influences can be either instantaneous or lagged, with a maximum time-lag of τ_{max}:

$$\mathbf{X}_t = \sum_{i=0}^{\tau_{max}} \mathbf{A_i X}_{t-i} + \eta_t \tag{6.6}$$

It can be rewritten as a vector autoregressive model without instantaneous effects:

$$\mathbf{X}_t = \sum_{i=1}^{\tau_{max}} \mathbf{M_i X}_{t-i} + \eta_t \tag{6.7}$$

This model, estimated through a least-square procedure, is used to obtain residuals of the prediction of $\mathbf{X}_t$. A standard LiNGAM analysis is then used on these residuals to obtain an instantaneous causal model A_0. Finally, $A_i, i > 0$ are deduced by a reparametrization of $M_i, i > 0$:

$$\mathbf{A}_i = (\mathbf{I} - \mathbf{A_0})\mathbf{M}_i, \forall_i, i \in (1, \ldots \tau_{max}) \tag{6.8}$$

6.4 Causal Discovery from Event Sequences

Multivariate time-series, represented as dynamic Bayesian networks, are one alternative to model temporal process, in which the value of each variable is measured at discrete time points and how these values are affected (caused) by previous values of the same or other variables are represented as causal arcs. A limitation of this type of representation is that the temporal intervals between time slices is considered fixed, so modeling a wide range of temporal interactions could result in a very complex model.

An alternative is to model the process as an *event sequence*. An event sequence with N events is denoted as $\{(t_i.e_i)\}_{i=1}^{N}$, where t_i is the timestamp of the ith event, and e_i is the corresponding event type. Causal discovery of event sequences aims at obtaining a causal graph relating event types.

As an example [2], consider a series of events related to a car accident. Assume that at time $t = 0$, an automobile accident occurs, that is, a *Collision*. A possible consequence for the person involved in the collision is a *Head Injury* which occurs almost immediatly. If a head injury occurs, the brain will start to swell and if left unchecked the swelling will cause *Dilated Pupils* within 0 to 5 min. A head injury also tends to cause *Unstable Vital Signs*, taking from 0 to 10 min to make them unstable. Figure 6.7 shows the events data across time, and Fig. 6.8 the corresponding causal graph.

Similarly to causal discovery from multivariate time-series, the methods for causal discovery for event sequences are divided in three main types: (i) Granger causality,

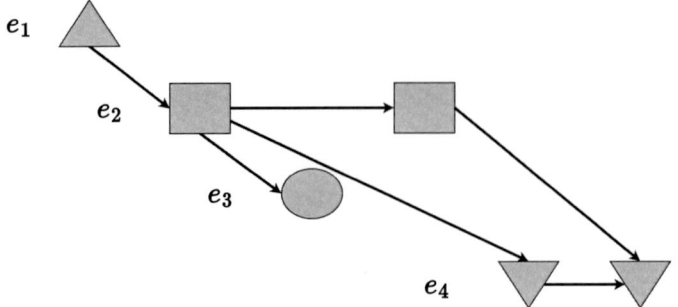

Fig. 6.7 Events for the car accident example: e_1, Collision; e_2, Head Injury e_3, Dilated Pupils; e_4, Unstable Vital Signs

Fig. 6.8 Causal graph for the car accident example: e_1, Collision; e_2, Head Injury e_3, Dilated Pupils; e_4, Unstable Vital Signs

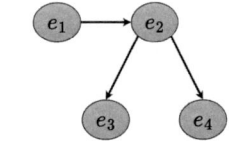

(ii) constraint-based, and (iii) score-based. Most of these methods are based on a modeling framework known as *Multivariate Point Processes*.

6.4.1 Multivariate Point Processes

A multivariate point process (MPP) refers to a stochastic process consisting of a series of binary events that occur in continuous time. MPPs are characterized by their high dimensionality, allowing for the inclusion of multiple types of events. A MPP with E types of events can be represented by a counting processes $\{N_e\}_{e=1}^E$, where $Ne = \{N_e(t), t \in [0, T]\}$ and the occurring time of these events $\{t_1, t_2, ..., t_n\}$ can be unevenly distributed.

MPPs are characterized by their conditional intensity function, which is defined as the expected instantaneous rate of occurrence for events of type e, given the historical information:

$$\lambda_e(t) = \frac{E[dN_e(t)|H_t]}{dt} \tag{6.9}$$

where $E[.]$ is the expected value, $\mathcal{E}$ represents all types of events, and $H_t = \{(t_i, e_i), t_i < t, e_i \in \mathcal{E}\}$ represents all types of events that happened before time t.

$\lambda(t)\Delta t$ represents the instantaneous probability of events of type e occuring in the time window $[t, t + \Delta t]$, which aligns to causal strength when condition on other variables. Thus, the causal discovery problem can be formulated as the task of taking a collection of point processes, where each individual process represents a sequence of events, and generating a causal graph G constructed from these processes. Within the causal graph G, each node corresponds to a point process, while each directed edge represents a causal interaction from one point process to another.

6.4.2   Causal Discovery Methods

The methods for causal discovery from event sequences are analogous to those for multivariate time-series. Here we present a summary of the main types of approaches, and refer the interested reader to the following review for more details [11], and further references.

6.4.2.1   Granger Causality

In a similar way as for multivariate time-series, the Granger causality of events of type e_j upon events of type e_i, is established by the predictive utility of $e_j(t), t < t_0$, in forecasting $e_i(t)$.

When considering the specifications of the model, the methodologies can be classified into three main categories: Hawkes process-based, Wold process-based, and neural point process-based. *Hawkes processes* exhibit temporal dynamics characterized by either intensities that increase abruptly or decay gradually. A *Wold process* is a class of point process defined in terms of a Markovian joint distribution of inter-event times, and is used to model causality in event sequences with low complexity. *Neural Point Processes* (NPPs) take advantage of the learning capabilities of neural networks to effectively model sequences, often surpassing statistical point processes in terms of predictive performance. NPPs encode the event sequence into a hidden state, capturing its underlying features, and subsequently employ decoders to infer the future intensity function.

6.4.2.2   Constraint-Based

Constraint-based methods rely on the assessment of conditional independence between processes, in a similar way as the examination of conditional independence (D-separation) in causal discovery for multivariate time series. Corresponding notions and procedures can be identified in constraint-based methods when applied to event sequences.

6.4.2.3   Score-Based

Score-based methods are based on the inference of causality through the optimization of score functions. For instance, *Score-Based Proximal Graphical Event Model* employs a score-based optimization approach to estimate the model parameters. It utilizes BIC as the score to search for optimal time windows and parent sets for each event type. In a similar approach, *CIR* (Conditional Intensity Ratios) introduces a collection of scores, such as conditional intensity-based scores, along with associated algorithms, to estimate the cause-effect associations between event pairs derived from extensive event datasets.

6.5 Subsampling

One of the main challenges of using data collected in time series is that causal interactions may occur on a time scale faster than the frequency of measurement [13]. This can lead to a loss of valuable information to determine the true causal relationships between events. An example of this can be seen in Fig. 6.9, where the original causal structure of the time series is shown (6.9a); and the causal structure of the same process under subsampling, obtained by making observations every two time steps (6.9b). If it is assumed that the structure of 6.9b is correct, valuable information about the true causal relationships between the variables can be lost. This may lead to believe that variable Z can be intervened to control Y, but the true influence of Z on Y is mediated by X. In this way, an intervention in X would be more effective. Similarly, if the structure of Figure 6.9b is used, the predictions of the behavior of the variables can be completely different from those obtained if the true causal structure of the time series is used.

6.5.1 Finding an Equivalence Class of Causal Structures

Given a causal graph obtained a certain measurement time scale that is known to be slower that the causal time scale (undersampling), it is possible to obtain a set of

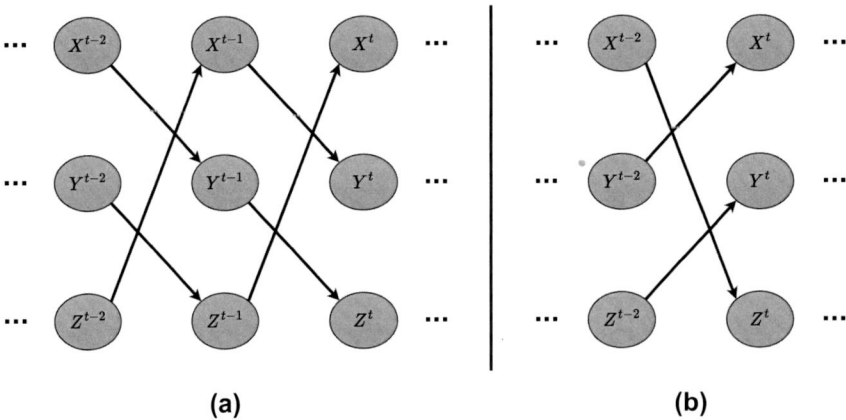

Fig. 6.9 a A causal model represented as a DBN. **b** The causal structure recovered from data with subsampling, obtained by making observations every two time steps

possible causal structures so that the *true* causal graph is consistent with one of the graphs in the set [6].

Plis et al. [19] developed an algorithm that, given the undersampled causal graph, it allows learning a set of causal structures assuming the level of subsampling is known. This is performed through a graphical representation of the causal structure of the time series. Subsequently, all the possible causal structures are obtained, comparing them with the measurement causal structure, which may be affected by some degree of sub-sampling. In this way, if the new structures are consistent with the measured structure they are considered as a possible causal structures, obtaining an equivalence class of causal structures.

The method is based on a search procedure for finding all consistent paths for each arc in the undersampled graph. Consider a sub-sampling rate of two, $u = 2$, and the causal graph obtained, G_2; the true graph is G_1. Given the rate of subsampling, an arc in G_2 represents a path in which there is an intermediate unknown variable in the intermediate time step that is not measured. For example, an arc $x_{t-1} \rightarrow y_t$ in G_2, implies a possible path $x_{t-2} \rightarrow v_{t-1} \rightarrow y_t$ in G_1, where v is a virtual node which could correspond to any variable in the graph. The method finds all possible consistent variables with v, and repeats this for all the arcs in G_2.

The use of this approach present various challenges that limit its use. As more nodes are present in the causal structure, the variations or possible structures that can be derived grows exponentially, so the computational resources required increase significantly. Furthermore, since the causal structure is obtained from the time series data, statistical errors may occur, which would imply that some structures are not consistent with the original causal structure, or that structures that are actually consistent are not taken into account. Hyttinen et al. [13] and Abavisani et al. [1] extend the previous approach by proposing a constraint satisfaction procedure which is computationally more efficient, and can also recover from conflicts due to statistical errors. An example of this method is illustrated in Fig. 6.10, considering a sub-sampling rate of two. The figure depicts the structure obtained from the subsampled data (top) and all the possible consistent structures (bottom). However, in the worst case finding the graphs that are consistent with the graph obtained or estimated at the measurement timescale is an *NP-Complete* problem.

6.5.2   Recovering the Structure of a Subsampled Time Series

A method to recover the *true* causal structure from a subsampled times series was proposed by [17]. It consists on *imputing*[1] the missing data using adversarial neural networks to try to recover the true causal structure. The trained model is fed with the subsampled time series in order to generate data that behaves similarly to the original time series, so the original causal structure can be recovered. The completed data series is then fed to a causal discovery algorithm.

---

[1] Imputation is the process of replacing missing data with substituted values.

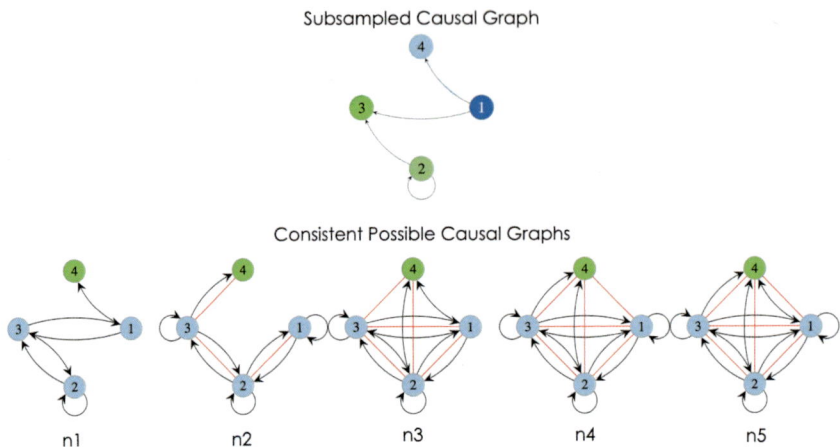

Fig. 6.10 Representation of the causal structure obtained from the subsampled data (top), and the structures, $n1, ..., n5$, that are consistent given the graph from the subsampled data (bottom)

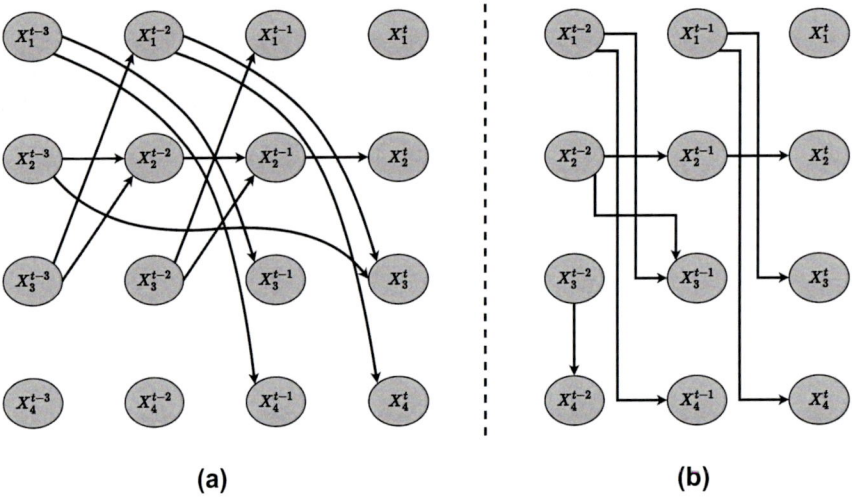

(a)   **(b)**

Fig. 6.11 a The original causal structure. **b** The causal structure of the subsampled time series

For data generation and imputation an adversarial neural network (ANN) [26] is used. This type of model involves the learning of regular patterns from the input data in such a way that the model can generate output data that may have a similar behavior, such that the generated data may be considered as part of the original data set. In this way, we may capture the original distribution by making the distribution of the outputs approximate the original data distribution. This is achieved by using two models: the generator that is trained to generate data sets based on the original data set, and the discriminator that aims to classify the received data as real or fake.

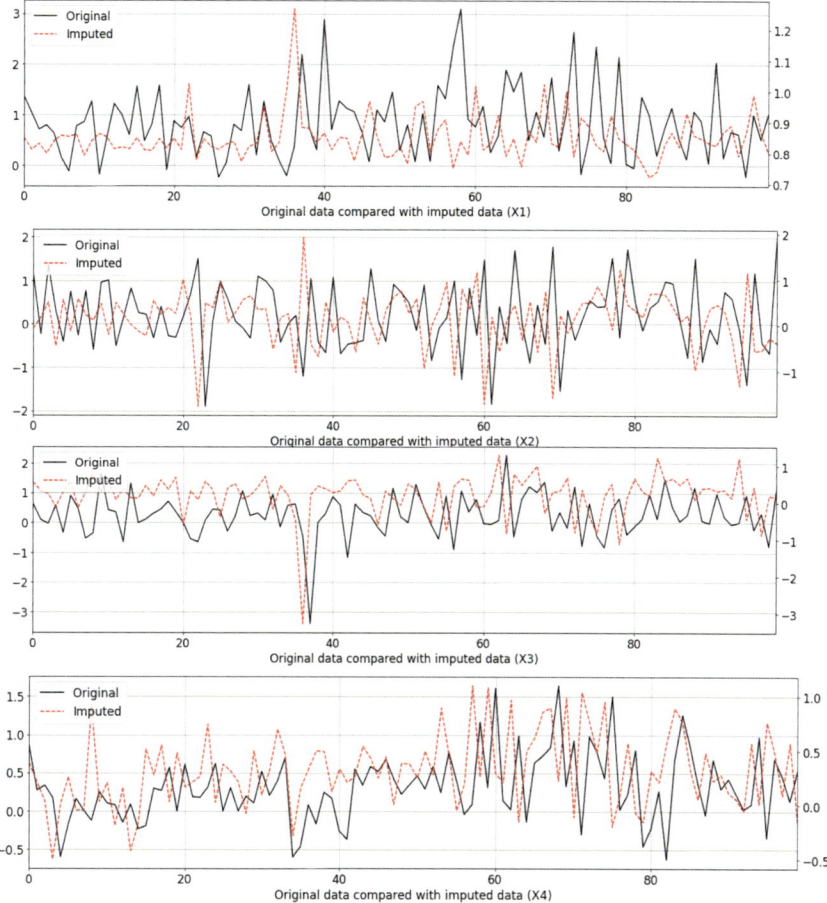

Fig. 6.12 Graphs of the original (black) and generated (red) data for variables (from top to bottom), X_1, X_2, X_3 and X_4. The generated data for each one of the variables resembles the behavior of the original values

These two models work together until the discriminator model accepts the generated data as if these data belong to the original data set.

Once the data generated complements the time series, the PCMCI algorithm (see Sect. 6.3.2.1) is used to reconstruct its causal structure. To illustrate the method we consider a time series with four variables as shown in Fig. 6.11. A time series composed of 10,000 observations was generated for each one of the variables. This same time series was affected by a subsampling rate of two, that is, the observed data comprise values of the variables every two time steps. Figure 6.11a depicts the causal structure of the original time series, and Fig. 6.11b the structure obtained from the subsampled data.

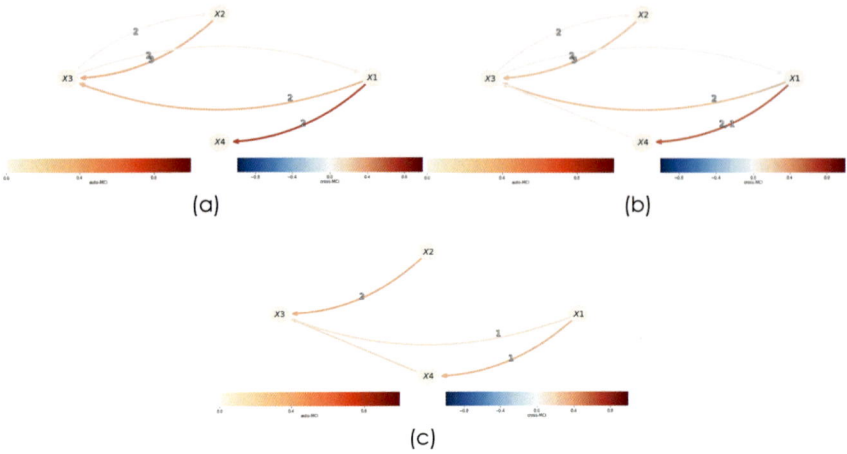

Fig. 6.13 Representation of the causal structure for: **a** the original time series, **b** the time series with imputed data, and **c** the subsampled time series. Each graph represents the causal structure via a compact representation (rolled graph), indicating the time delay (number associated to the link) and strength (color code) of each causal link

The time series was completed with imputed data for each one of the observed variables: X_1, X_2, X_3 and X_4 using the ANN previously trained. A comparison is depicted in Fig. 6.12, where it can be seen that the generated data is very close to real data, presenting a similar behavior over time. The mean absolute error (MAE), which is the measure of the difference between both sets of values, allows to quantify the precision of the generated values compared to the original values, resulting in a value of 0.019 (approx. 2%). Then, the PCMCI algorithm was applied to the complemented data, obtaining a very similar structure to the original one as shown in Fig. 6.13.

6.6 Additional Reading

A comprehensive survey of causal discovery methods from time series data is presented in [3]. A more recent survey is described in [11], which includes causal discovery from event sequences. The PCMCI method is described in detail in [20,21]. Algorithms for estimating the cause-effect association between pairs of events from event datasets is described in [5]. The problem of undersampling and different schemes to cope with it are included in [1,13,17,24]. A dataset for causal discovery from time series is introduced in [16].

6.7 Exercises

1. Given the dynamic Bayesian network in Fig. 6.1, specify the required conditional probability tables assuming time-invariance and binary variables.
2. Represent the causal structure in Fig. 6.11a as a rolled graph.
3. What will be the resulting structure of the model in Fig. 6.1 if this will be learned from a time series with a subsampling rate of 2?
4. Consider a time series with three variables, X_1, X_2, X_3. After testing for predicting each variable with past observations we obtained the following results: X_1 can be predicted based on X_1 with a time lag of 1 and X_3 with a time lag of 2; X_2 is predicted based on X_1 with a time lag of 2 and X_3 with a time lag of 1; and X_3 can be predicted with X_3 with a time lag of 1. What will be the structure of the time series based on this tests: (a) represented as a series of times slices, (b) represented as a rolled graph.
5. Given the dataset in Table 6.2, estimate the causal structure based on Granger causality using the test of prediction improvement, with a maximum time lag of 1.
6. Repeat the previous problem using the test for conditional mutual information.
7. Estimate the parameters from the structure obtained in the previous problem and the data in Table 6.2; for this exercise discretize the data in 5 equal width intervals.
8. Given the data set in Table 6.2, apply the first phases of the PCMCI algorithm: (a) build a partially connected graph considering a maximum time lag of 2, draw the resulting structure; (b) test for conditional independence as in the PC algorithm (assume that the process is stationary) and eliminate the arcs according to the algorithm (define a threshold), draw the resulting structure.
9. *** Implement the PCMCI algorithm and apply it to the data in Table 6.2, as well as other data sets.
10. *** Investigate the algorithm to recover the set of possible structures given a subsampled time series [1, 13, 24] and implement it; apply it to the structure in Fig. 6.11 to obtain all possible consistent structures.

Acknowledgements The section on causal discovery from event sequences is based on [11], and the section on undersampling is inspired by the work of Muñoz [17]. Section 6.2 is based on [25], used with permission from Springer.

Table 6.2 Data for the exercises. t indicates the time index of the data for X_1 and X_2

t	X1	X2	t	X1	X2
1	2.0	0.7	26	3.6	1.8
2	2.1	1.4	27	1.4	1.2
3	1.1	1.3	28	0.5	−0.3
4	1.1	1.3	29	1.1	1.3
5	1.2	0.3	30	0.7	1.1
6	2.3	1.2	31	1.5	−0.1
7	0.8	0.1	32	1.0	0.0
8	−0.9	0.4	33	−1.0	0.6
9	0.1	1.2	34	1.7	−0.4
10	1.2	−0.3	35	0.1	−1.1
11	1.0	−0.3	36	0.4	−1.5
12	−0.3	−3.3	37	−1.3	−2.7
13	−2.1	0.6	38	−1.7	1.6
14	−0.1	1.2	39	−0.8	3.3
15	0.3	2.1	40	0.7	1.3
16	0.0	1.8	41	0.8	1.5
17	1.3	1.0	42	1.4	0.2
18	0.9	0.1	43	1.2	1.1
19	3.0	−0.5	44	2.9	−1.0
20	−1.5	0.6	45	0.3	3.0
21	−0.5	0.2	46	−2.0	0.6
22	−1.0	2.0	47	−1.3	1.2
23	1.1	0.3	48	1.6	1.4
24	0.9	0.4	49	−0.8	−0.4
25	−0.4	−0.9	50	0.6	−2.0

References

1. Abavisani M, Danks D, Plis S (2023) GRACE-c: generalized rate agnostic causal estimation via constraints. In: The eleventh international conference on learning representations
2. Arroyo-Figueroa G, Sucar LE (1999) A temporal bayesian network for diagnosis and prediction. In: Proceedings of the 15th conference on uncertainty in artificial intelligence (UAI). Morgan-Kaufmann, San Mateo, pp 13–20
3. Assaad CK, Devijver E, Gaussier E (2022) Survey and evaluation of causal discovery methods for time series. J Artif Intell Res 73:767–819
4. Baum CF, Hurn S, Otero J (2022) Testing for time-varying Granger causality. Stata J Promoting Commun Stat Stata 22(2):355–378
5. Bhattacharjya D, Gao T, Mattei N, Subramanian D (2020) Cause-effect association between event pairs in event datasets. In: Proceedings of the twenty-ninth international joint conference on artificial intelligence (IJCAI-20), pp 1202–1208
6. Danks D, Plis S (2014) Learning causal structure from undersampled time series. iN: JMLR: workshop and conference proceedings

7. Friedman N, Murphy K, Russell S (1998) Learning the structure of dynamic probabilistic networks. In: Proceedings of the fourteenth conference on uncertainty in artificial (UAI). Morgan Kaufmann Publishers Inc., pp 139–147

8. Gerhardus A, Runge J (2020) High-recall causal discovery for autocorrelated time series with latent confounders. In: 34th conference on neural information processing systems (NeurIPS2020). Vancouver, Canada

9. Geweke J (1982) Measurement of linear dependence and feedback between multiple time series. J Am Stat Assoc 77(378):304–313

10. Granger C (1969) Investigating causal relations by econometric models and cross-spectral methods. Econometrica 37(3):424–438

11. Gong C, Zhang C, Yao D, Bi J, Li W, Xu Y (2024) Causal discovery from temporal data: an overview and new perspectives. ACM Comput Surv 57(4)

12. Hacker RS, Hatemi-j A (2006) Tests for causality between integrated variables using asymptotic and bootstrap distributions: theory and application. Appl Econ 38(13):1489–1500

13. Hyttinen A, Plis S, Jarvisalo M, Eberhardt F, Danks D (2017) A constraint optimization approach to causal discovery from subsampled time series data. Int J Approximate Reasoning 90:208–225

14. Hyvärinen A, Zhang K, Shimizu S, Hoyer PO (2010) Estimation of a structural vector autoregression model using non-gaussianity. J Mach Learn Res 11:1709–1731

15. Lutkepohl H (2005) New introduction to multiple time series analysis. Springer Science and Business Media

16. Munoz-Benítez J, Enrique Sucar L (2023) Synthetic time series: a dataset for causal discovery. CLeaR Datasets Track. https://www.cclear.cc/2023/AcceptedDatasets/munozbenitez23a.pdf

17. Munoz-Benítez J, Enrique Sucar L (2023) Data imputation with adversarial neural networks for causal discovery from subsampled time series. MICAI LNAI 14392:39–51

18. Pamfil R, Sriwattanaworachai N, Desai S, Pilgerstorfer P, Georgatzis K, Beaumont P, Aragam B (2020) Dynotears: structure learning from time-series data. In: Chiappa S, Calandra R (eds) Proceedings of the twenty third international conference on artificial intelligence and statistics, volume 108 of proceedings of machine learning research, pp 1595–1605

19. Plis S, Danks D, Yang J (2015) Mesochronal structure learning. In: Proceedings of uncertainty in artificial intelligence

20. Runge J (2018) Causal network reconstruction from time series: from theoretical assumptions to practical estimation. Chaos: Interdisc J Nonlinear Sci 28(7)

21. Runge J (2020) Discovering contemporaneous and lagged causal relations in autocorrelated non-linear time series datasets. In: Peters J, Sontag D (eds) Proceedings of machine learning research, vol 124, pp 1388–1397

22. Sanchez-Romero R, Ramsey JD, Zhang K, Glymour MRK, Huang B, Glymour C (2019) Estimating feedforward and feedback effective connections from fMRI time series: assessments of statistical methods. Netw Neurosci 3(2):274–306

23. Shojaie A, Fox EB (2022) Granger causality: a review and recent advances. Ann Rev Stat Appl 9(1):289–319

24. Solovyeva K, Danks D, Abavisani M, Plis S (2023) Causal learning through deliberate undersampling. In: 2nd conference on causal learning and reasoning

25. Sucar LE (2021) Probabilistic graphical models: principles and applications. Springer Nature, Switzerland

26. Yoon J, Jarrett D, Van der Schaar M (2019) Time-series generative adversarial networks

Causal Reinforcement Learning

7

Abstract

This chapter introduces *causal reinforcement learning*, a recent field that combines reinforcement learning and causal discovery. After a brief review of Markov decision processes and reinforcement learning, we present (i) how a causal model can be used to accelerate learning an optimal policy, and (ii) how to learn and use at the same time a causal model in the process of learning an optimal policy.

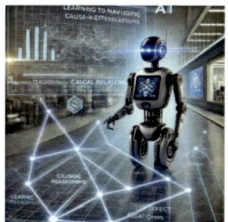

7.1 Introduction

Reinforcement Learning (RL) allows agents to explore their environments, learning how to solve certain task through the rewards associated with actions taken at each state encountered along the way. The goal for these agents is to determine the optimal policy for each state, i.e., the best action to take at each state. In recent years, we have observed a notable boost in the field, driven by the integration of RL with deep neural networks, known as *deep reinforcement learning* (DRL). Despite the notable advancements, there are many unresolved issues. One of the main limitations of existing approaches is the failure to exploit patterns beyond the associative, which hinders agents from effectively understanding and utilizing causal relationships and

© The Author(s), under exclusive license to Springer Nature Switzerland AG 2026   141
L. E. Sucar, *Causal Discovery*, Computer Science Foundations and Applied Logic,
https://doi.org/10.1007/978-3-031-98345-0_7

latent structures present in the environment. This, on one side, implies that it takes a large number of trials to learn an optimal policy in complex problems; and on the other side, the learned policies can not be easily transferred to other similar domains.

Causal Reinforcement Learning (CRL) is an emerging field where two essential areas for the development of artificial intelligence are integrated. Existing works in the area have shown how causality can contribute to mitigate some of the limitations of reinforcement learning [3], ranging from data-inefficiency, lack of interpretability, and long learning times, among others. Other works have demonstrated that it is not necessary to have a causal model *a priori*; but that a causal model and an optimal policy can be learned *simultaneously*, such that each one benefits the other.

RL is in certain way related to causal reasoning and discovery, as performing actions in the world is essentially the same as doing interventions. However, until recently, both areas have been treated as separate fields. The integration of both fields provides several advantages to both: (i) having a causal model of the environment can help to learn faster an optimal policy, and can provide the agent with the capacity to explain its decisions and transfer them to other similar problems, and (ii) reinforcement learning provides an alternative to learn causal models, as the agent is doing interventions, and thus provides additional useful data that that goes beyond only learning from observational data.

Next we present a brief overview of reinforcement learning, including an introduction to Markov decision processes. Then, we describe some approaches for using a given causal model to improve reinforcement learning, in particular to accelerate the learning process. Finally, we end the chapter we with the most challenging problem, on how an RL agent can learn and at the same time use a causal model.

7.2 Reinforcement Learning

Reinforcement learning (RL) is an area of machine learning concerned with how agents can learn to make good sequences of decisions. The following definition is given by Sutton and Barto [10]: "Reinforcement Learning is learning what to do, how to map situations to actions so as to maximize a numerical reward signal. The learner is not told which actions to take, as in most forms of machine learning, but instead must discover which actions yield the most reward by trying them. In the most interesting and challenging cases, actions may affect not only the immediate reward but also the next situation and, through that, all subsequent rewards. These two characteristics, trial and error search and delayed reward, are the two most important distinguishing features of reinforcement learning". The formal definition of reinforcement learning algorithms is based on the assumption that the environment has the Markov property, an environment that satisfies the Markov property is called a Markov decision process.

7.2.1 Markov Decision Processes

A Markov decision process (MDP) [9] models a sequential decision problem, in which a system evolves over time and is controlled by an agent. The system dynamics are governed by a probabilistic transition function Φ that maps states S and actions A to new states S'. At each time, an agent receives a reward R that depends on the current state s and the applied action a. By solving an MDP representation of the problem, we obtain a *policy* that maximizes the expected reward over time and that also deals with the uncertainty of the effects of an action.

For example, consider an agent (a simulated robot) that *lives* in a grid world, so the state of the robot is determined by the cell where it is; see Fig. 7.1. The robot wants to go to the goal (cell with a smiling face) and avoid the obstacles and dangers (filled cells and a forbidden sign). The robot's possible actions are to move to the neighboring cells (up, down, left, right). We assume that the robot receives a certain immediate reward when it passes through each cell, for instance $+50$ when it arrives to the goal, -50 if it goes to the forbidden cell (this could represent a dangerous place), and -1 for all the other cells (this will motivate the robot to find the shortest route to the goal).

MDPs consider that there is uncertainty in the result of each action. For example, if the selected action is *up* the robot goes to the upper cell with a probability of 0.8 and with probability of 0.2 to other cells. If there is also uncertainty about the state of the agent, the problem becomes a *Partially Observable* MDP (POMDP). The objective of the robot is to go to the goal cell as fast as possible and avoid the dangerous states. This will be achieved by solving the MDP that represents this problem, and maximizing the expected reward. The solution will provide the agent with a policy, that is, what is the best action to perform in each state,

Formally, an MDP is a tuple $M = < S, A, \Phi, R >$, where S is a finite set of states $\{s_1, \ldots, s_n\}$. A is a finite set of actions $\{a_1, \ldots, a_m\}$. $\Phi : A \times S \times S \rightarrow [0, 1]$ is the state transition function specified as a probability distribution. The probability of reaching state s' by performing action a in state s is written as $\Phi(a, s, s')$. $R : S \times A \rightarrow \mathbb{R}$ is the reward function. $R(s, a)$ is the reward that the agent receives if it takes action a in state s.

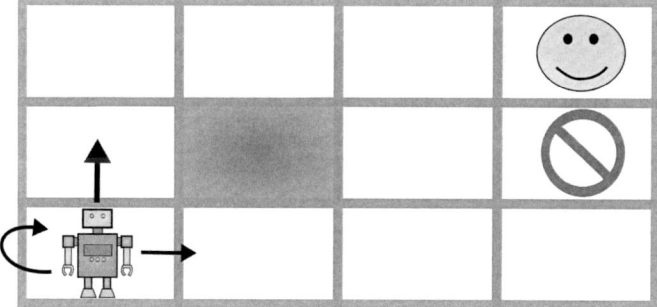

Fig. 7.1 A robot in the grid world. Cells represent the possible states of the robot, with a smiling face for the goal and a forbidden sign for danger. The robot is shown in a cell with the width of the arrows illustrating the probability for the next state given the action *up*

A *policy*, π, for an MDP is a function $\pi : S \to A$ that specifies for each state, s_i, the action to be executed, a_i. Given a certain policy π, the expected accumulated reward for following π from state s, is known as the *value* for that state according to the policy, $V^\pi(s)$. For the infinite horizon case, a parameter known as the *discount factor*, $0 \le \gamma < 1$, is included so that the sum converges. This parameter can be interpreted as giving more value to the rewards obtained at the present time than those obtained in the future. Including the discount factor, the value function is written as:

$$V^\pi(s) = R(s, a) + \gamma \sum_{s' \in S} \Phi(a, s, s') V^\pi(s') \tag{7.1}$$

For the discounted infinite-horizon case with any given discount factor γ, there is a policy π^* that is optimal regardless of the starting state and that satisfies what is known as the *Bellman* equation:

$$V^\pi(s) = max_a\{R(s, a) + \gamma \sum_{s' \in S} \Phi(a, s, s') V^\pi(s')\} \tag{7.2}$$

The policy that maximizes the previous equation is then the optimal policy, π^*:

$$\pi^*(s) = argmax_a\{R(s, a) + \gamma \sum_{s' \in S} \Phi(a, s, s') V^\pi(s')\} \tag{7.3}$$

Given that the model is known (transition and reward functions), there are three basic methods for solving an MDP and finding an optimal policy: (a) value iteration, (b) policy iteration, and (c) linear programming. Next we describe the value iteration algorithm, for the alternative algorithms see the additional reading section.

Value iteration consists in iteratively estimating the value for each state, s. Note that this is actually a set of N equations, one for each state, $s_1, s_2, \ldots, s_N$. It starts by assigning an initial value to each state; usually this value is the immediate reward for that state. Then, these estimates of the values are improved in each iteration through a maximization process using the *Bellman* equation. The process is terminated when the value for all states *converges*, this is when the difference between the values in the previous and current iterations is less than a predefined threshold. The actions selected in the last iteration correspond to the optimal policy. The method is shown in Algorithm 7.1.

Algorithm 7.1 The Value Iteration Algorithm

1: $\forall_s V_0(s) = R(s, a)$ {Initialization}
2: $t = 1$
3: **repeat**

4:     $\forall_s V_t(s) = max_a\{R(s, a) + \gamma \sum_{s' \in S} \Phi(a, s, s') V_{t-1}(s')\}$ {Iterative improvement}

5: **until** $\forall_s \mid V_t(s) - V_{t-1}(s) \mid < \varepsilon$
6: $\pi^*(s) = argmax_a\{R(s, a) + \gamma \sum_{s' \in S} \Phi(a, s, s') V_t(s')\}$ {Obtain optimal policy}

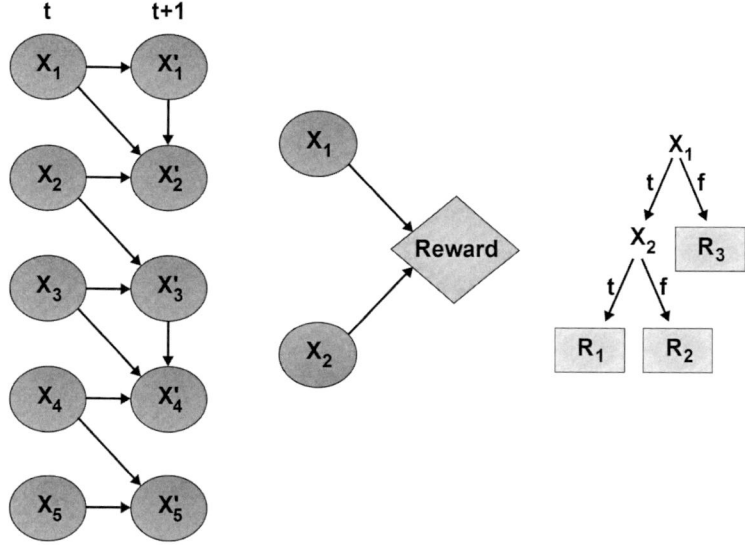

Fig. 7.2 Left: A DBN with 5 state variables that represents the transition function for one action. Center: Influence diagram denoting a reward function. Right: Structured conditional reward (CR) represented as a binary decision tree

If the model is not known, the alternative is to explore the environment to learn by trail and error the optimal policy, using reinforcement learning. Before we go into reinforcement learning, we will describe a more compact and intuitive representation of an MDP which is useful for causal RL.

7.2.2   Factored MDPs

Solving very *large* MDPs could become problematic. By a large MDP we refer to an environment with a very large number of states or actions. For instance, robot navigation in a large space represented as a grid in which each cell is a state, could have hundreds of thousands of states. A manipulator with many degrees of freedom where each one can be controlled, could imply a large number of possible actions. An alternative is to decompose the state space and take advantage of the independence relations to reduce the memory and computation requirements, using a graphical model-based representation of MDPs known as *Factored* MDPs.

In a factored MDP, the set of states is described via a set of random variables $\mathbf{X} = \{X_1, \ldots, X_n\}$, where each X_i takes on values in some finite domain $Dom(X_i)$. A state s defines a value $x_i \in Dom(X_i)$ for each variable X_i. The transition model and reward function can become exponentially large if they are explicitly represented as matrices, however, the frameworks of dynamic Bayesian networks and decision trees give us the tools to describe the transition model and the reward function concisely.

Let X_i denote a variable at the current time and X'_i the variable at the next step. The transition function for each action, a, is represented as a two–stage dynamic Bayesian network, that is a two–layer directed acyclic graph G_T whose nodes are $\{X_1, \ldots, X_n, X'_1, \ldots, X'_n\}$; see Fig. 7.2 (left). Each node X'_i is associated with a *conditional probability distribution* $P_\Phi(X'_i \mid Parents(X'_i))$, which is usually represented by a matrix (*conditional probability table*) or more compactly by a decision tree. The transition probability $\Phi(a, s_i, s'_i)$ is then defined to be $\Pi_i P_\Phi(X'_i \mid \mathbf{u_i})$ where $\mathbf{u_i}$ represents the values of the variables in $Parents(X'_i)$.[1]

The reward associated with a state often depends only on the values of certain features of the state. The relationship between rewards and state variables can be represented with value nodes in an influence diagram, as shown in Fig. 7.2 (center). The conditional reward table (CRT) for such node is a table that associates a reward with every combination of values for its parents in the graph. This table is exponential in the number of relevant variables. Although in the worst case the CRT will take exponential space to store the reward function, in many cases the reward function exhibits structure allowing it to be represented compactly using decision trees or graphs, as shown in Fig. 7.2 (right).

7.2.3 Reinforcement Learning in Discrete Spaces

First we describe RL considering discrete state and actions spaces; in the next section we consider the problem of a continuous state space.

In RL, we consider that the learner (agent) and environment interact during a sequence of time steps, $t = 0, 1, 2, 3, \ldots$. At each time t, the agent receives some representation of the environment's state, $s_t \in S$, where S is the set of possible states, and based on its knowledge, selects an action, $a_t \in A(s)$, where $A(s)$ is the set of actions available in state s. One time step later, as a consequence of its action, the agent receives a numerical reward, $R(s_t, a_t)$, and finds itself in a new state, s_{t+1}, see Fig. 7.3. At each time step, the agent implements a mapping from states to probabilities of selecting each possible action. This mapping is called the agent's policy and is denoted by π, where $\pi(a \mid s)$ is the probability of taking action a at state s. The agent's goal is to maximize the total amount of reward it receives over the long run. There are different reward models, the most common is the total expected discounted reward:

$$R_t = \sum_{t=0}^{\infty} \gamma^t r_{t+1} \tag{7.4}$$

where is $0 \leq \gamma < 1$ is the discount factor.

---

[1] In this case, *synchronic* arcs (that connect variables in the same time step) are not considered for time t, and could be included or not at time $t + 1$.

Fig. 7.3 In the RL context, an agent interacts with its environment executing actions a_t according to its current policy, and receiving the state information s_{t+1} and certain reward r_t

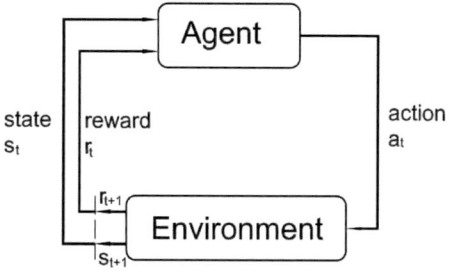

There are several algorithms to learn the optimal policy, here we describe one of the simplest, *Q-learning*. In this approach, an estimated action-value function, $Q : S \times A \rightarrow R$ is learned, where $Q(s, a)$ is the expected return found when executing action a from state s, and greedily following the current policy thereafter. The current best policy is generated from Q by simply selecting the action that has the highest value for the current state. *Exploration*, when the agent chooses an action to learn more about the environment, must be balanced with *exploitation*, when the agent selects what it believes to be the best action. One simple approach that balances the two is the $\varepsilon-$greedy action selection, where the agent selects a random action with small probability ε, and the current best action is selected with probability $1 - \varepsilon$. Q-learning is summarized in Algorithm 7.2; α is the *learning rate* and γ the discount factor. The process is repeated for a number of episodes and at the end the action that gives the highest Q value for each state provides the *best* policy. Under certain conditions the algorithm will converge to the optimal policy, although it could take a high number of episodes (long time).

Algorithm 7.2 The Q-learning Algorithm.

1: $\forall_{s,a} Q_0(s, a) = R(s, a)$ {Initialization}
2: **repeat**

3:     Initialize s
4:     **repeat**

5:         Select a in s according to $\varepsilon-$greedy
6:         $Q(s, a) \leftarrow Q(s, a) + \alpha[r + \gamma max_a Q(s', a') - Q(s, a)]$
7:         $s \leftarrow s'$

8:     **until** s is a terminal state

9: **until** Last episode

Q-learning is a *model-free* algorithm. A model-free algorithm is an algorithm that estimates the optimal policy without using or estimating the dynamics (transition and reward functions) of the environment. A *model-based* algorithm uses the transition function (and the reward function) in order to estimate the optimal policy. The agent might have access only to an approximation of the transition function and reward functions, which can be learned by the agent while it interacts with the environment or it can be given to the agent (e.g., by another agent). In a model-based algorithm, the agent can predict the dynamics of the environment because it has an estimate of the transition function, and use it to accelerate learning. However, note that the transition and reward functions that the agent uses in order to improve its estimate of the optimal policy might just be approximations of the "true" functions, and in some cases this could be detrimental to learning the optimal policy. For more information consult [10] and see the additional reading section.

7.2.4 Reinforcement Learning in Continuous Spaces

As the state space grows, using a discrete representation of the state space becomes impractical, or impossible if the state space is continuous. In such cases, RL learning methods use function approximmators, such as artificial neural networks, which rely on parameterized functions and use supervised learning methods to set their parameters. Function approximation is used in large or continuous tasks to better generalize experience. Parameters and biases in the approximmator are used to abstract the state space so that observed data can influence a region of state space, rather than just a single state, and can substantially increase the speed of learning.

In RL for large state spaces, function approximmators represented as neural networks are used to approximate the Q values, so it is known as *deep reinforcement learning* (DRL). The neural network is parameterized by a set of weights, W, and it has $|A|$ outputs, each one approximates the Q value for an action a, where A is the set of possible actions. The input to the neural network is a feature representation X_t of the state s_t at time t. Thus, the neural network computes a function $F(X_t, W, a)$ to approximate $Q(s_t, a)$.

For example, consider the robot navigation of Fig. 7.1, but instead of having a small grid, we have a very large grid or even a continuos state space. The robot state corresponds to its position on the environment, which could be represented by a set of features such as those obtained from a camera in the robot that observes the surrounding environment. Considering 4 actions as in the grid world example, the neural network will map the features from the current state to the Q values for each of the four actions, as shown in Fig. 7.4.

The weights $\mathbf{W}$ of the neural network have to be learned via training. However, in principle we need to know the values of the outputs, $Q(s, a)$, to calculate a loss function in terms of the difference with the estimated values, but we do not know these values at the current time. A bootstrapping trick is used to estimate the loss

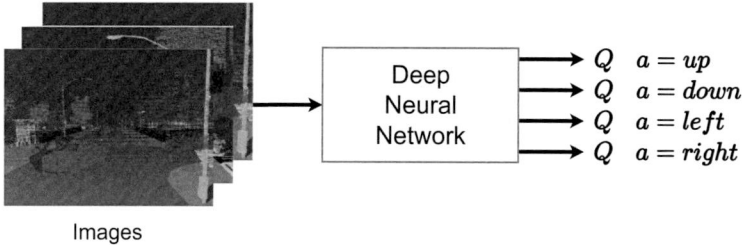

Images

Fig. 7.4 The robot navigation example in the context of deep RL. Based on the features extracted from the images that represent the robot state (current position in the world), the Q value for each action is estimated by the deep neural network

function based on an approximation of the Q-values in the future. This surrogate observed values are obtained by the Bellman equation:

$$Q(s_t, a_t) = r_t + \gamma max_a \hat{Q}(s_{t+1}, a) \tag{7.5}$$

Essentially, we are looking at all actions one step ahead to obtain an improved estimate of $Q(s_t, a_t)$. For this we must wait to observe the next state and reward, by performing the action a_t to calculate $\hat{Q}(s_{t+1}, a)$. And now, we can obtain the loss function and update the weights in the neural networks by backpropagation.

The previous steps of prediction and learning are performed simultaneously, as the values of actions are used to update the weights and select the next action, until reaching a terminal state for which the value must be set to zero. As in the case of a discrete state space, a balance between exploration and exploitation must be considered to select the next action, using, for instance, ε-greedy. The action with the largest predicted Q value is selected with probability $1 - \varepsilon$; otherwise, a random action is selected.

We can consider two variants of the previous method: *off-policy* and *on-policy*. The previously described approach, known as Q-learning, is an *off-policy* method; as the best possible action at each step is used to update the parameters, that might be different for the action actually executed. In contrast, in *on-policy* approaches such as *SARSA* [10], the action actually selected is used to perform the update.

There are basically three main variants of the DRL techniques:

1. *Critic-only methods.* These techniques, such as Q-learning and SARSA, are based on a value function (i.e., Q) that is optimized. This function is a critic, and the policy is derived from this critic.
2. *Actor-only methods.* An alternative are policy-gradient methods, that directly learn the probability of the different actions, and the advantages of the actions are estimated using Monte Carlo sampling. In this case, an NN is trained to predict the probability of the various actions. Based on these probabilities, a biased random selection of the action to be executed is performed. Once the action is executed the

reward is observed, and this is used to improve the weights in the NN to maximize the expected reward.

3. *Actor-critic methods*. These methods combine both previous approaches. They have two coupled NNs: a policy network and a Q-network. The policy network learns the probabilities of the different actions; and the Q-network learns their Q-values, in order to provide an estimation to the policy network. Thus, the policy network uses the estimated Q values to weight its updates of the corresponding NN. The Q-network is updated using an on-policy approach, and the policy (action to be executed) is controlled by the policy network.

For more information of the different DRL algorithms see the Additional Reading section.

7.3 Using a Causal Model for Reinforcement Learning

Reinforcement learning performance can be improved with causal knowledge as side information, assuming that the causal information is known and given *a priori*. The causal information is used for different purposes in different scenarios: (i) to accelerate learning the optimal policy, (ii) to deal with latent confounders in different settings like Multi-Armed Bandits (MAB) and Markov Decision Processes (MDPs), (iii) for off-policy evaluation to mitigate heterogeneity and data scarcity, or (iv) to derive causal explanations about the behavior of model-free RL agents. Next we will focus on the first case, using prior causal knowledge to speed up learning a policy in RL in scenarios that can be modeled as goal-conditioned Markov decision processes.

In a goal-conditioned Markov decision process, a task has an underlying causal structure that describes the behavior of the environment. The state space is represented as a factored-MDP by a set of state variables, S. In some cases, the state representation is at a low-level of abstraction which it is not useful for representing and using a causal model; so it can be transformed to a set of higher-level variables, X. It is assumed that a causal graphical model is known, or at least its structure, given by G, representing the causal relations for (some) variables in X.

According to the above considerations, the problem can be represented as a tuple $(S, A, X, \Phi, R, G, \psi)$, where:

- S is the state space,
- A is the set of possible actions,
- X is the abstract state representation,
- $\phi : S \times A \times S' \to [0, 1]$ is the transition function that defines the probability of the next state s' given that the action a is executed in state s,
- $R : S \times A \to \mathbb{R}$ defines the immediate reward r that the agent receives when action a is executed in state s,
- G is a causal graph that rules the environment,
- $\psi : S \to X$ is a function that associates a state s to an abstract state x.

In the case of a continuous state space, it is assumed that there is a function ψ that maps the states to the abstract variables X.

7.3.1   Causal-Based Q-Learning

Given a causal model, we can modify the Q-learning algorithm by adding an extra option, so besides selecting the best action according to Q or executing a random action (ε-greedy), it might opt to query the causal model [2]. In the latter case, the agent analyses the causal model and determines if certain possible action given the current state, can take it closer to the goal. If such action is available it selects it; otherwise it applies the traditional ε-greedy exploration strategy.

When using the causal model, the agent exploits the following query: What action(s) leads to meet the current goal? The variables in the causal model, G, relevant to the current goal, g, are defined as the set E. So the agent must determine which action(s) affect the variables in E. This means that performing an action, a, causes one or more variables in E to change their value, bringing the agent closer to its goal. This strategy narrows the space of possible actions, i.e., the agent prefers those actions that bring it closer to the goal and at the same time it does not change those variables that have the desired value. By computing the set of predecessors of the goal according to the causal structure, the agent selects relevant actions to reach the goal; and this will help to learn faster the optimal policy.

The set E is obtained through a function that calculates which variables have a different value between the goal g and the vector of abstract variables X. In some cases, the variables of interest may follow a causal order, which means that one goal depends on other subgoals. Therefore, the list E can be stored in a data structure such that its elements are ordered by a priority function. Thus, the agent chooses first those actions that lead to the final goal, and then those actions that affect the variables that are direct predecessors of the goal (parents in the causal graph), and so on.

For example, consider the classical RL problem of a taxi that has to pick up a passenger at certain location and deliver her to the destination location.[2] A partial causal model for this case is shown in Fig. 7.5, including some of the abstract state variables and some actions. The ordered list of variables in this case will be:

$$\mathbf{E} = [X_4, X_2, X_3, X_1] \tag{7.6}$$

So the agent will prefer first to execute an action that makes X_4 true (passenger delivered), second the action that affects X_2, etc. Of course not all the actions could be possible at certain state, so the agents selects them based on this order from the possible actions.

Causal-based Q-learning is depicted in Algorithm 7.3.

---

[2] The Taxi task is described in detail in the next Section.

Algorithm 7.3 The Causal-Based Q-learning Algorithm

1: $\forall_{s,a} Q_0(s, a) = R(s, a)$ {Initialization}
2: **repeat**

3:    Initialize s
4:    **repeat**

5:       $x \leftarrow \psi(s)$ {ψ transforms the state s to the abstract state x}
6:       $E \leftarrow f(g, x)$ {f gets a list of abstract variables in the causal model that have a value different from the desired one}
7:       **if** $E \neq \emptyset$ **then**

8:          Select a such that it is a predecessor of $x \in E$ according to the priority

9:       **else**

10:          Select a in s according to ε−greedy

11:       **end if**
12:       $Q(s, a) \leftarrow Q(s, a) + \alpha[r + \gamma max_a Q(s', a') - Q(s, a)]$
13:       $s \leftarrow s'$

14:    **until** s is a terminal state

15: **until** Last episode

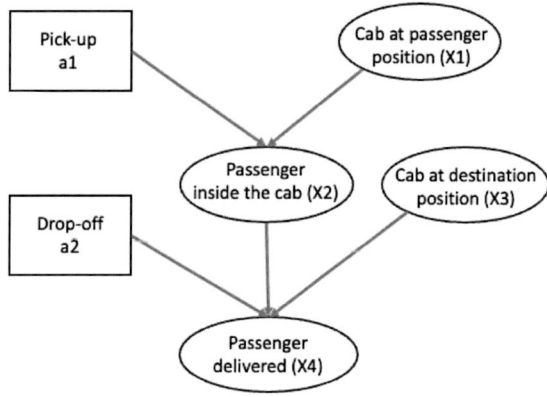

Fig. 7.5 A partial graphical causal model for the taxi environment. Action variables are depicted as rectangles and abstract state variables as ovals

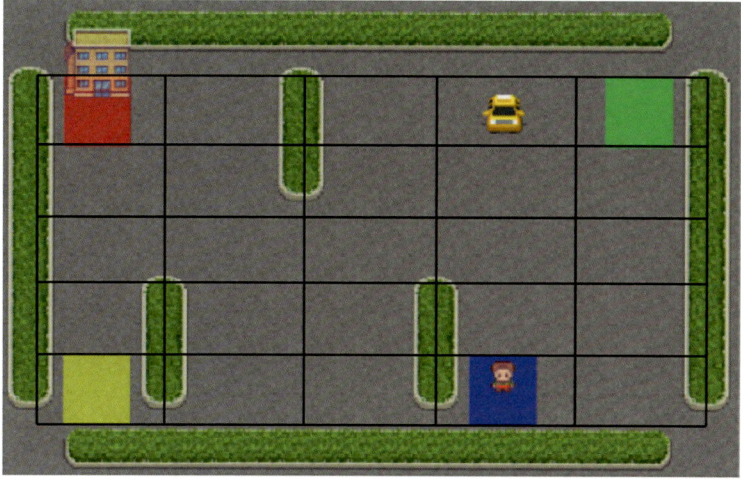

Fig. 7.6   The Taxi task environment. The grid world size is 5 × 5. There are four possible locations for the passenger pick-up and drop-off, marked using colored squares

7.3.1.1   An Example: The Taxi Problem

The Taxi task is a classical RL problem, in which a taxi in a grid world has to pick up a passenger at certain location and deliver him to another location. In a 5 × 5 grid, there are four relevant positions in the world labeled R, B, G, Y. In each episode, the cab starts at a randomly chosen square. There is a passenger at one of the four positions (also chosen randomly), and the passenger wishes to be transported to one of three remaining positions. The cab must go to the passenger's position, pick him up, go to his destination and drop him off. The episode ends when the passenger is dropped off at its destination. The actions set A is composed of six elements: one action to pick up the passenger, $a1$, one to drop off the passenger, $a2$, and four navigation actions that move the cab one frame north, south, east or west, denoted by $a3, a4, a5, a6$, respectively. There is a reward of –1 for each action, an additional reward of 20 for each passenger successfully brought to the destination, and a penalty of –10 for illegal pickup and drop-off actions. The state space S has 500 three–element tuples: the 25 squares, the 5 passenger positions (including when in the cab), and the 4 destinations. The taxi environment is illustrated in Fig. 7.6.

In this case, it is required to transform the state variables to a more convenient representation for the causal model. So the set S is transformed to the set X, that contains 4 variables that translate the tuples with the cab and passenger positions into binary variables:

- x_1 it expresses whether the cab is in the same position as the passenger,
- x_2 denotes whether the passenger is taken inside the cab,
- x_3 describes whether the cab is in the target position, and
- x_4 represents the state that the passenger is delivered correctly.

It is a relatively simple task, so the goals are $G = g_1, g_2$, where $g_1 = [1, 1, 0, 0]$ and $g_2 = [1, 1, 1, 1, 1]$, which correspond to the values of x_1, x_2, x_3, x_4, respectively. The first vector can be seen as the subgoal (the passenger boards the cab) and the second vector is the overall goal (to deliver the passenger to his destination). The causal graph that represents the relations between these abstract state variables and some of the action variables is depicted in Fig. 7.5.

To evaluate the impact of using the causal model, the causal-based Q-learning algorithm was compared to the traditional Q-learning algorithm [2]. Both algorithms are tested on two versions of the environment, one deterministic and the other stochastic. The deterministic version corresponds to the one where the agent successfully executes the action it has selected. In the stochastic environment, attempting a certain action does not imply that it will be carried out successfully. This behavior is simulated by assigning a high probability to correctly executing the selected action and with a low probability, spread evenly over the rest of the actions. For example, it may be that when attempting to drop off the passenger, the taxi moves northward.

As usually done when evaluating RL algorithms, the experiment was repeated M times, in this case 50, to have statistically significant results. Each repetition consists of 1000 episodes, and each episode of 200 steps of the RL algorithms. For a fair comparison, both algorithms used the same parameters:

- Learning rate, $\alpha = 0.8$.
- Discount factor, $\gamma = 0.95$.
- Exploration parameter, $\varepsilon_{max} = 1.0$ and $\varepsilon_{min} = 0.1$ (it is linearly decreased from ε_{max} to ε_{min}).
- For the stochastic case, probability of executing the selected action, $P_a = 0.8$; and $P_o = 0.2$ of executing another action, randomly selected.

The results are summarized in Fig. 7.7 for the deterministic environment and in Fig. 7.8 for the stochastic environment. Each graph depicts the number of episodes in the horizontal axis, and, in the vertical axis, the average reward (and standard deviation) obtained by the agent in the 50 repetitions.

These results show the benefit of using causal knowledge in the process of learning a policy in RL. In both cases, it can be observed that the causal-based agent starts at a relatively high average reward, so the causal model helps to achieve relatively good results almost without learning (this is known as *jump start*). Also, it converges earlier to the maximum average reward, approx. 200 episodes before in the deterministic environment and 50 in the stochastic one. Thus, even in this relatively simple problem there is a significant benefit of using a causal model, so we can expect at even larger benefit in more complex tasks.

It has been shown that even a partial, or incorrect causal model, can help to accelerate learning [3], which is important if the agent does not have a causal model initially, and has to learn it during the process of learning the policy. This is the topic of the next section.

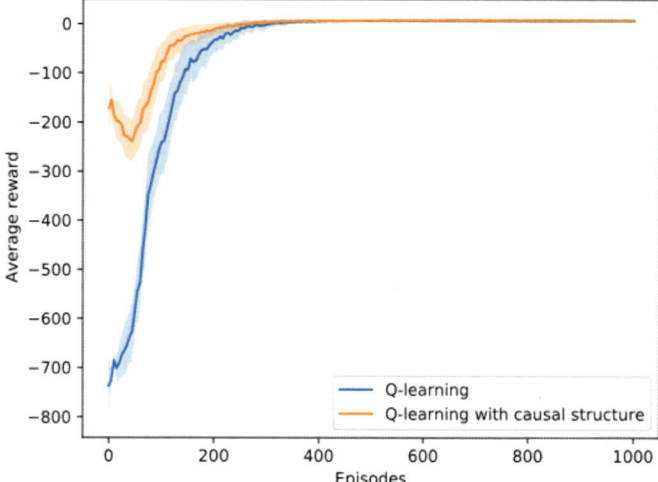

Fig. 7.7 Performance comparison between Q-learning and causal-based Q-learning in the deterministic environment. Each line corresponds to the average reward in 50 repetitions and the shaded region is the standard deviation

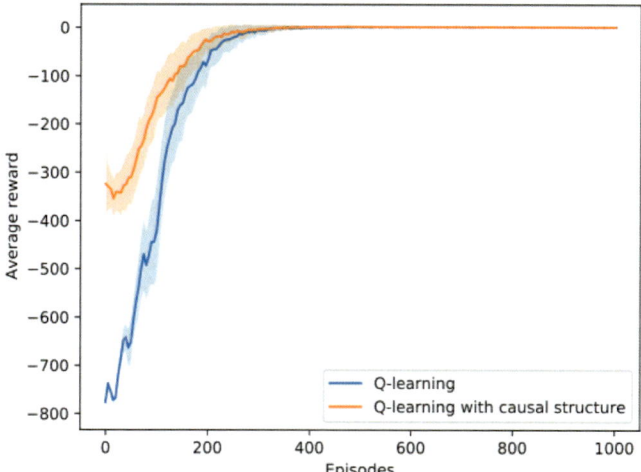

Fig. 7.8 Performance comparison between Q-learning and causal-based Q-learning in the stochastic environment. Each line corresponds to the average reward in 50 repetitions and the shaded region is the standard deviation

7.4 Learning and Using a Causal Model During Reinforcement Learning

In some domains it could be difficult to provide a causal model a priori, so it is desirable that the agent can also learn a causal model in the process of reinforcement learning, a more challenging task. RL methods allow for sampling the environment as opposed to learning from a static data set of pre-collected observations (like in traditional causal discovery scenarios). This interactive learning setting for data collection can be beneficial for online decisions, e.g. deciding on which variable to intervene based on how informative an intervention is for estimating a causal structure. The data collected are biased by the feedback cycles between the environment, the agent and the action selection policy. That is, given a sequence of actions of the agent on the environment, it will only be possible to obtain observations of the sequence of states and rewards that the environment returns to the agent given its intrinsic dynamics. However, two advantages for causal discovery are: (i) to consider the data's temporal order, and (ii) its interventional nature. An analogous challenge to traditional RL appears: exploration, which is convenient for causal discovery, versus exploitation, which is good for policy learning.

In [5], a framework which simultaneously learns and uses causal models for efficient induction of causality and task policies in online MDP settings is proposed, named CARL (Causality-Aware Reinforcement Learning). In a synergistic way, CARL alternates between three stages: (i) (RL for CD), where it promotes the selection of actions to obtain better causal models in fewer episodes than traditional methods of obtaining data in RL, (ii) (CD), where a score-based algorithm is used to learn causal models and (iii) (RL using CD), where the learned models are used to select actions that speed up the learning of the optimal policy by reducing the number of interactions with the environment.

7.4.1 CARL

Figure 7.9 illustrates the CARL framework. The interaction between the agent and the environment is controlled by the combination algorithm, which tells the agent what stage to perform according to a predefined combination strategy. During the RL for CD stages, the agent employs the Causality-Concerned Action Selection (CCAS) algorithm to choose actions that potentially give more useful information for causal discovery. In CD stages the agent applies a causal discovery algorithm on the collected dataset to discover the causal model represented as a set of Dynamic Bayesian Networks (DBNs), one per action. In the RL using CD stages, the agent uses the Causal-Based Action Selection (CBAS) algorithm to filter the set of possible actions to keep those with the highest expected immediate reward, based on estimations from a series of simulated interventions in the learned causal model. Once an action is executed in the environment, the agent receives the reward and transits to a new state, and the process is repeated until a maximum number of time steps has been reached.

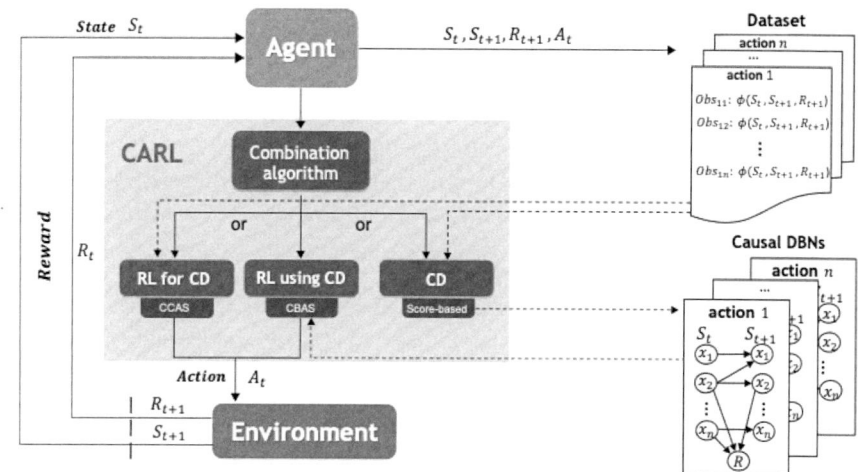

Fig. 7.9 The CARL framework. The agent learns at the same time a policy and a causal model by alternating between: RL for CD, an exploration strategy to help learning the causal model; RL using CD, an exploration strategy that takes advantage of the causal model learned so far; and CD, which uses the data from the previous explorations to update the causal model. Figure from [5]

The backbone of CARL is the combination algorithm, where the agent does not know anything in advance more than the variables conforming the state. So, the agent, which is trying to learn the optimal policy for the given task but also to discover the underlying causal structure, interacts with the environment according to a combination strategy. That strategy tells the agent what to execute for the following T episodes. Two combination strategies are proposed. The first one is $C1 = [$RL for CD, CD, RL using CD, CD$]*$ (The * means that the strategy is repeated until the end of the learning episode). This strategy essentially consists of evenly alternating the RL for CD and RL using CD stages throughout the training process while doing CD in between with the collected data. In this way, the agent is giving the same importance to learning the causal models as to learning the optimal policy. The second strategy is $C2 = [$RL for CD, CD, RL using CD$]$, RL for CD is performed only at the beginning of training, the models are learned and then used until the end.

It is important to mention that neither the best action for causal discovery (RL for CD) nor the action with the best immediate reward according to the causal model (RL using CD) has to correspond to the best long-term action for the agent. For that reason, to secure the convergence of the algorithm to the optimal policy, there is a probability of $1 - \varepsilon$ that our agent ignores the action indicated by these two exploration strategies, and selects the action which maximizes the value function $Q(s, a)$. As in traditional RL, the exploration factor should be decreased during training, so that in the final episodes the agent mostly exploits to adjust the value function and converges to the optimal policy.

The different stages of the method are detailed next.

7.4.1.1 Exploration for Causal Discovery

Normally, in MDP settings, the agent acts in the environment according to an exploration strategy such as epsilon-greedy. The agent can not directly intervene on any state variable, it can only visit states that are the result of the implicit transition mechanism in the environment after taking the sequence of actions guided by the policy. Therefore, to estimate the interventional causal model, the data will be heavily influenced by the frequency at which the policy suggests performing each action. However, if we want to learn causal models, then we would prefer to have more balanced data that leads to better models.

A new stage called (RL for CD) is introduced in which the agent is concerned with trying to select actions that favor causal discovery rather than learning the optimal policy. Basically, the agent keeps a record of the number of times it has performed each action in each state. Then, if it is time to explore, instead of performing a random action as traditional epsilon-greedy, the agent first tries to select the less explored action in the given state to obtain more balanced datasets for each of the actions and increase the probability of learning a correct causal model per action. When all actions in a given state are executed at least certain number of times, random action selection is started. In practice, a limit between 10 and 100 executions off each action per state provides a good compromise.

7.4.1.2 Causal Discovery

During the exploration stages, the agent stores the following data for each executed action: the state variables at time t, $\mathbf{X_t}$, the state variables at time $t + 1$, $\mathbf{X_{t+1}}$, and the reward, r. Thus, $|\mathbf{A}|$ data sets are obtained, on per action, with n_i tuples, where n_i is the number of times that action a_i was executed. To identify the causal models, various causal discovery algorithms can be utilized, which may be further enhanced by imposing additional constraints: (i) no time $t + 1$ variable can cause a variable at time t, and (ii) variables at the same time point cannot cause one another. Given the structural constraints, the causal discovery process becomes easier because learning the skeleton of the causal model is equivalent to learning its full structure.

In this case, a causal model per action is learned, represented as a two-stage dynamic Bayesian network, see Sect. 7.2.2; so the agent has to learn $|\mathbf{A}|$ causal models. Later we will show these causal models for the Taxi task example.

To guarantee that the learned skeleton fully corresponds to the ground truth causal model, a generative mechanism that allows the agent to set each relational state variables X_i to all its possible values before taking an action and sampling the transition and reward values is required. In the RL context, it is not possible to intervene all the state variables, only those corresponding to the actions, so it can not be guaranteed that the learned models are complete. However, it has been shown [3] that partially correct causal models are enough to speed up policy learning. As more data is collected in the exploration stages, this can be used to improve the causal models.

7.4.1.3   Reinforcement Learning Using Causal Models

As described in Sect. 7.3, a causal model can help to accelerate learning an optimal policy in RL. In this work, the pervious approach is adapted to consider causal models represented as two-stage DBNs, and extended to take into account not only the actions to obtain an immediate positive reward, but also to filter those actions that can lead to an undesirable large negative reward.

In this stage, the agent uses an epsilon-greedy exploration strategy. When the agent explores (does not select the best action), for each of the possible actions a_i, the agent calculates the probability distribution for the reward variable given a representation of the state $\mathbf{X}$ and the causal model learned so far for that action. Inference is performed by taking into account only the variables that are parents of the immediate reward variable in the corresponding learned causal model. In deterministic environments, those parent variables can be only a subset of the state variables at time t, but in the stochastic environments the reward can depend also on state variables at time $t + 1$, so the inference must be performed in two steps. First, variables $\mathbf{X_{t+1}}$ must be inferred using $\mathbf{X_t}$, and then r is inferred using both $\mathbf{X_t}$ and $\mathbf{X_{t+1}}$. If there is a probability greater than a threshold value of obtaining immediate positive reward by taking action a_i, this is selected; otherwise, if there is a probability greater than a threshold of obtaining high negative reward by executing action a_i, action a_i is discarded from the set of possible actions in the current state. An intuitive alternative to setting the threshold value could be to initialize it at a small value and gradually increase it, since the quality of the models is expected to improve as the episodes progress.

7.4.1.4   An Example: The Taxi Task Revisited

To exemplify the CARL method, we consider again the Taxi Task. However, now the agent does not know the causal models *a priori*, it has to learn them simultaneously as it learns the optimal policy.

The first experiment is designed to measure the effectiveness of the action selection strategy (RL for CD) to improve causal discovery, compared against an agent performing exploration using the traditional epsilon-greedy exploration (this alternative agent is named as PGM-22, [4]). The PGM-22 agent learns and uses causal models, but in the RL stage it does not select actions for causal discovery purposes, it uses epsilon-greedy instead. The CARL agent, on the other hand, uses the proposed combination strategy $C1 = $ [RL for CD, CD, RL using CD, CD]* where the (RL) stage is replaced by the (RL for CD) stage.

An important parameter in both strategies is T, indicating how often the causal discovery is carried out, so different values of T are used: 10, 20, 50, 100, to measure the effect of using RL for CD stage vs RL independently of T. In all experiments, the exploration rate (ε) is linearly decreased from 1.0 to 0.1 over 1000 episodes. For each value of T the experiment was repeated 10 times.

To evaluate the quality of the discovered causal model, the *structural Hamming distance* (SHD) is used. This distance represents the minimum number of edge changes required (insertions, deletions, and inversions) to transform one model into

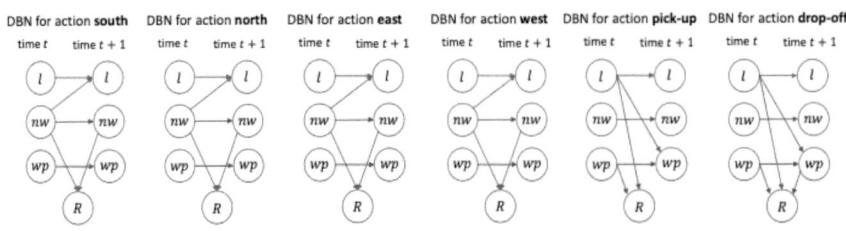

Fig. 7.10 Ground truth causal DBNs, one per action, relating abstract state variables at consecutive time steps, t and $t + 1$, and reward, R, for each action in the Taxi task stochastic environment. The deterministic case is equivalent but removing all edges from variables at $t + 1$ to R. The variable l represents whether the cab is in the road, at the passanger origin or destination, wp denotes whether the passenger is taken inside the cab, nw indicates in which direction is the nearest wall

another. A value of 0 indicates that the discovered model is equal to the ground truth. As there is a causal model for each action, the reported SHD in a given episode is the sum of the SHDs among all discovered models for each agent. As mentioned previously, the causal knowledge is represented as a set of causal graphical models, one per action. The ground truth models for the Taxi task are shown in Fig. 7.10 (these are not known by the agent, they are just used to evaluate the learned models).

In Fig. 7.11 we can see the results of these experiments for different values of the T parameter in the deterministic version of the Taxi task. We can see in all cases that the agent using the RL for CD stage (CARL) manages to discover better models in fewer episodes than the agent that uses the traditional exploration strategy (PGM-22).

In the second experiment, CARL is compared against traditional model-free (Q-Learning) and model-based (Dyna-Q) algorithms in terms of policy learning, considering also the PGM-22 algorithm (kind of an earlier version of CARL). Dyna-Q is a model-based reinforcement learning algorithm that extends Q-learning by incorporating an estimated model of the environment that it uses to plan ahead N steps in imagination and update the value function accordingly. The objective is to evaluate if CARL is able to learn an optimal policy faster than traditional algorithms while discovering the underlying causal models in the process. It was tested with different values of the exploration rate (epsilon), since this parameter has a significant influence on the performance of the traditional RL algorithms. This parameter starts with a high value ($\varepsilon = 1.0, 0.7$), linearly decreasing the value to 0.1. The experiment is repeated for 20 trials for each algorithm, with 1000 episodes per trial, reporting the average cumulative reward (avg-reward) with its standard deviation along the episodes.

The experiment was done for both strategies: $C1 = $ [RL for CD, CD, RL using CD, CD]*, that alternates RL for CD and RL using CD stages while doing CD in between; and $C2 = $ [RL for CD, CD, RL using CD], in which RL for CD is performed only at the beginning. Figure 7.12 summarizes the results for both strategies and for two different initial values of ε.

For strategy $C1$ (Fig. 7.12 left), we can see that CARL obtains significantly higher rewards in episodes where it uses the model (RL using CD) while its performance

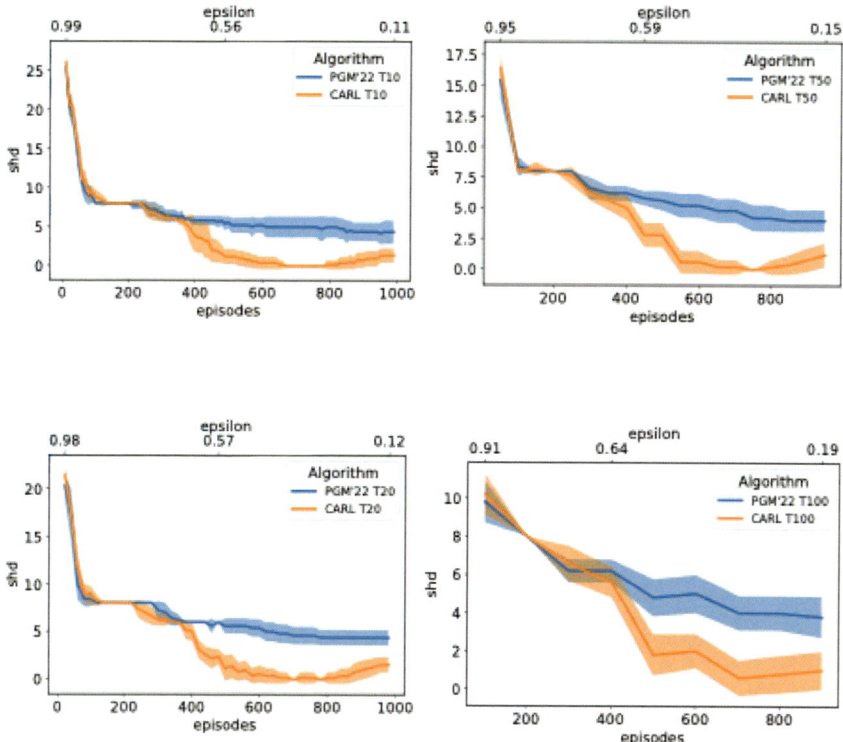

Fig. 7.11 Causal discovery results in the Taxi task comparing an agent that does traditional exploration (PGM-22) versus an agent that uses the exploration strategy oriented for causal discovery (CARL). The graphs depict the structural Hamming distance (lower is better) against the number of episodes for different values of T (10, 20, 50, 100)

either drops or equals that of other algorithms when it performs actions to improve the model (RL for CD). The difference is proportional to the exploration factor, as a higher exploration rate favors causal discovery, and having *good* causal models has in impact in learning the optimal policy when these models are used.

An interesting observation from previous experiments is that the RL algorithms obtain better results when it starts using the discovered causal models, even if they are incomplete. In the next experiments, we evaluate what would happen if we only perform the (RL for CD) stage once and subsequently learn and use the discovered models until the end of the episodes, using strategy $C2$. The results are shown in Fig. 7.12 right. As it can be seen, regardless of the exploration factor and despite the fact that the models used are not complete when first used, the method achieves a significant boost in reward per episode when starting to use the discovered models, which it maintains throughout the episodes, being able to learn the optimal policy in far fewer episodes than the other algorithms.

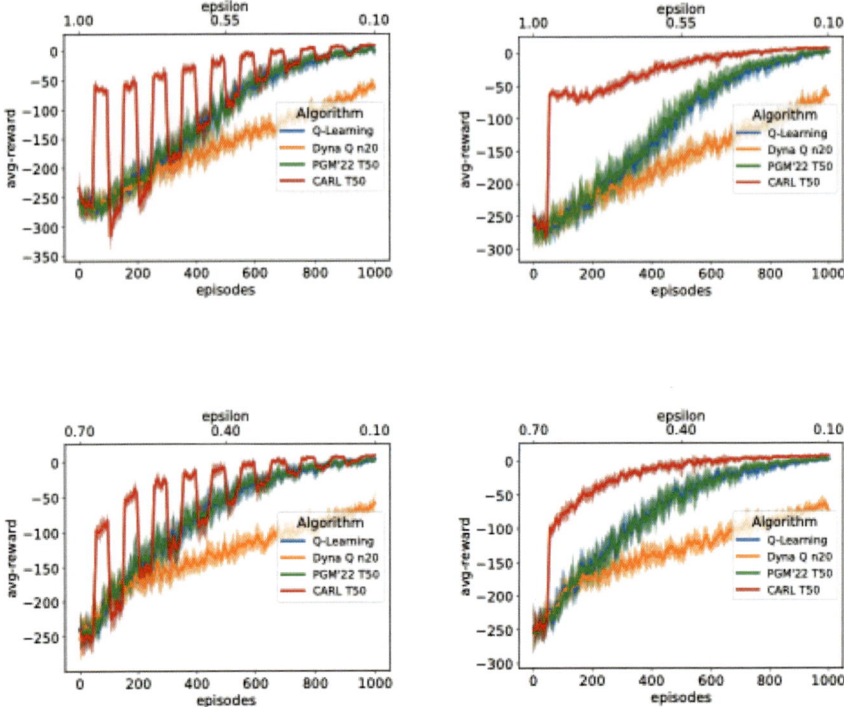

Fig. 7.12 Policy learning results in the Taxi task comparing four algorithms: Q-learning, Dyna-Q, PGM-22 and CARL. The two graphs on the left show the results for strategy $C1$, in which the CARL agent alternates between (RL for CD) and (RL using CD); and the two graphs to the right consider strategy $C2$, in which it does causal discovery only once and then (RL using CD). The graphs depict the average reward (higher is better) against the number of episodes for different values of ε, in the top graphs it starts at 1 and it is linearly decreased to 0.1, in the bottom it starts at 0.7 and it is decreased to 0.1. Figure from [5]

Comparing both strategies, $C1$ and $C2$, the second one tends to produce better results in policy learning, which are more stable. However, the disadvantage of $C2$ is that the causal models might never be fully discovered, while $C1$ tends to recover better models as it continues to perform exploration for CD (this was demonstrated experimentally, see [5]).

7.5 Generalizability

Generalizability refers to the ability of a trained policy to perform effectively in new and unseen scenarios. It poses a major challenge in the deployment of RL algorithms in real-world applications. Traditional RL algorithms are typically designed for solv-

Fig. 7.13 Results for the extended Taxi scenario. The graphs show the average rewards (and standard deviation, 20 trails) versus number of training episodes obtained by CARL using different percentages of the previously learned causal model and the standard Q-Learning algorithm without causal knowledge. Figure from [5]

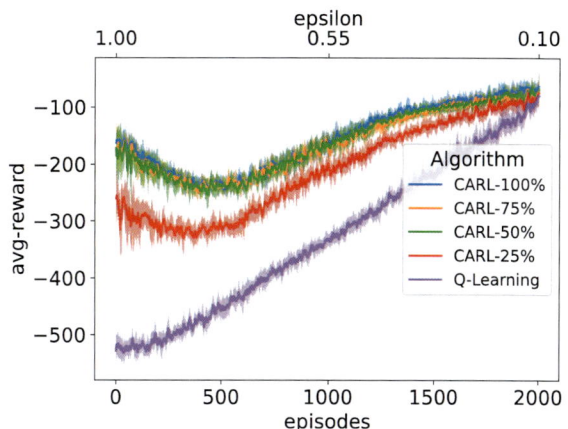

ing a single task, and they can easily overfit the environment, failing to adapt to minor changes.

Generalization involves various settings. Zero-shot generalization entails that the agent solely acquires knowledge in training environments and then this is applied to unseen scenarios. Alternatively, the agent receives additional training in the target domain, categorized as transfer learning.

Causal knowledge tends to be general, so it applies to different tasks and scenarios. This should allow for easy transfer of the learned knowledge to more complex scenarios that share the same set of relational variables and causal mechanisms.

To evaluate the generalization of CARL, it was tested in a variant of the Taxi scenario. This new environment doubles the size of the world and new walls are placed in different locations. In addition, the possible origin and destination points for the passenger are expanded. In this case the state space increases significantly, as the grid world size is 10×10 and the possible locations for the passenger and destination are increased to 6, so the size of the state space is 4200.

CARL was evaluated in this new scenario, using the causal knowledge acquired in the original taxi task. In this case, only the stage of CD for RL is applied utilizing the previously learned causal model; and it omits the stages of RL for CD and CD. To measure the impact of model quality on task learning, the agent starts with the full model (100%) but also utilizes partially correct models at different percent of correct edges with respect to the ground truth (75, 50, and 25%). It is also compared to the model-free Q-Learning algorithm. The results in terms of average rewards versus number of training episodes are depicted in Fig. 7.13.

From the results in Fig. 7.13, it is shown that CARL achieves certain zero-shot generalization, as it has a relatively high reward before it starts learning, in particular when it has the complete causal model. Regarding transfer learning, it can be noticed the CARL agent achieves much faster convergence than the model-free agent, even if the transferred models are incomplete. It seems that there are not significant differences between the full model (100%) and partial models at (75%) and (50%).

7.6   Additional Reading

Puterman [8] is an excellent reference for MDPs, including their formal definition and the different solution techniques. A review of MDPs and POMDPs is presented in [9]. The classical book on reinforcement learning is [10]. For deep reinforcement learning see [1]. The CARL method is described in detail in [5,6]. A survey on causal reinforcement learning is presented in [11].

7.7   Exercises

1. For the grid world example of Fig. 7.1, consider that each cell is a state and that there are four possible actions: *up, down, left, right*. Complete the specification of the MDP, including the transition and reward functions.
2. Solve the MDP for the previous exercise by value iteration. Initialize the values for each state to the immediate reward, and show how the values change with each iteration.
3. Define a factored representation for the robot in the grid world example, Fig. 7.1, considering that a state is represented in terms of two variables, *row* and *column*. Specify the transition function as a two-stage DBN and the reward function as a decision tree.
4. Given the grid world example of Fig. 7.1, the reward defined in Sect. 7.2.1, and four possible actions: *up, down, left, right*, simulate 10 iterations of the Q-learning algorithm considering $\alpha = 0.8$, $\gamma = 0.9$, and an ε−greedy action selection with $\varepsilon = 0.2$. Show the Q table after each iteration.
5. Given the light switches problem [7], see Fig. 7.14, define a graphical causal model for each alternative scenario (5 switches and 5 lights): (a) one to one, each switch controls on light, (b) common cause, a switch controls several lights (you can define how many), (c) common effect, a light is controlled by several switches (you can define how many).
6. For the previous problem, define the parameters of the causal model, for (a) a deterministic environment, (b) a stochastic environment.
7. Given the casual DBNs in Fig. 7.10, calculate the structural Hamming distance between the graphs for action *west* and action *pick-up*.
8. Propose an alternative strategy to combine RL for CD and CD for RL in the CARL algorithm.
9. *** Implement the causal-based Q-learning algorithms and apply it to the light switches problem.
10. *** Implement the CARL algorithm and test it in other RL problems.

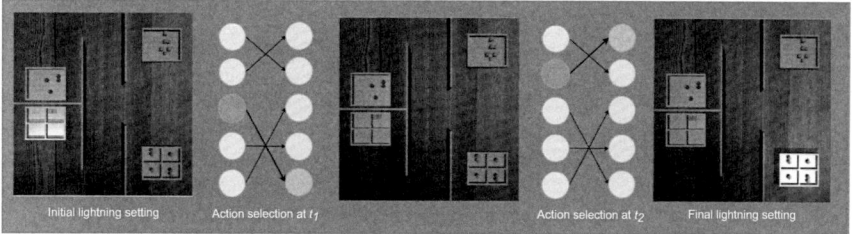

Fig. 7.14 The light switches problem. The agent has to turn ON/OFF the corresponding switch to turn on or off the desired lights. The graphs correspond to the one to one scenario, with the circles on the left representing the switches, and the circles on the right the lights

Acknowledgements Most of this chapter is inspired by the work of Mendez-Molina and Feliciano-Avelino [4–6]. Sections 7.2.1 and 7.2.2 are based on [9], used with permission from Springer. Figures 7.9, 7.12 (modified, only part of the figure is kept) and 7.13 are reproduced according to a Creative Commons License: https://creativecommons.org/licenses/by-nc-nd/4.0/.

References

1. Aggarwal CC (2023) Neural networks and deep learning. Springer Nature Switzerland
2. Feliciano-Avelino I, Méndez-Molina A, Morales EF, Sucar LE (2021) Causal based action selection policy for reinforcement learning. In: Proceedings of the Mexican international conference on artificial intelligence. Springer, pp 213–227
3. Méndez-Molina A, Feliciano-Avelino I, Morales EF, Sucar LE (2020) Causal based Q-learning. Res Comput Sci 149(3):95–104
4. Méndez-Molina A, Morales EF, Sucar LE (2022) Causal discovery and reinforcement learning: a synergistic integration. Int. Conf. Probabilistic Graph. Models (PGM). In: Salmerón A, Rumí R (eds) Proceedings of machine learning research, vol 186, pp 421–432
5. Méndez-Molina A, Morales EF, Sucar LE (2023) CARL: a synergistic framework for causal reinforcement learning. IEEE Access, vol 11
6. Méndez-Molina A (2024) CARL: A Synergistic Framework for Causal Reinforcement Learning. Ph.D. Thesis, Instituto Nacional de Astrofísica, Óptica y Electrónica, Puebla, Mexico
7. Nair S, Zhu Y, Savarese S, Fei-Fei L (2019) Causal induction from visual observations for goal directed tasks. arXiv:1910.01751
8. Puterman ML (1994) Markov decision processes: discrete stochastic dynamic programming. Wiley, New York
9. Sucar LE (2021) Probabilstic graphical models: principles and applications, 2nd edn. Springer
10. Sutton RS, Barto AG (1998) Reinforcement learning—an introduction, Adaptive computation and machine learning MIT Press
11. Zeng Y, Cai R, Sun F, Huang L, Hao Z (2024) A survey on causal reinforcement learning. IEEE Trans Neural Netw Learn Syst 1–21

Part III
Applications

In this third and last part, we will illustrate the application of causal discovery in different domains. In the first chapter, we will cover applications in biomedicine, the second one is dedicated social sciences, and the third one to artificial intelligence and robotics.

Applications in Biomedicine

8

Abstract

This chapter describes several applications of causal discovery in biomedicine, in three different domains: medical, neuroimaging and gene regulation. In the medical domain, we present a method for modeling the Attention Deficit Hyperactivity Disorder (ADHD); and a work that investigates the causal risk factors for COVID-19. Then, we describe two approaches to recover brain effective connectivity from different modalities of neuro images. Lastly, we include two examples of causal discovery of gene regulatory networks, one based on observational data, and other that combines observational and interventional data.

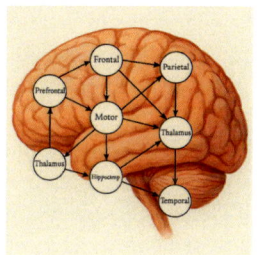

8.1 Introduction

Discovery of causal models from biological data is very important as these models help to obtain a deeper understanding of biological systems, which can be useful, for instance, for the treatment of certain diseases. However, this area is challenging, as in many cases it is not possible to perform experiments for ethical or practical reasons; and in the case of observational data, this tends to be quite limited.

© The Author(s), under exclusive license to Springer Nature Switzerland AG 2026   169
L. E. Sucar, *Causal Discovery*, Computer Science Foundations and Applied Logic,
https://doi.org/10.1007/978-3-031-98345-0_8

In this chapter, we provide examples of causal discovery in biomedicine in three different application domains:

- In a medical setting, regarding learning a causal model of the main variables that affect the attention deficit disorder (ADHD); and the causal risk factors for hospitalization and morbidity in the COVID pandemic.
- In neuroimaging, for learning the effective connectivity in the brain, that can be considered as causal relations between different brain regions. For this we present two application examples, one based on functional Near-Infrared spectroscopy (fNIRS) data and other using functional Magnetic Resonance Imaging (fMRI) data.
- In genetics, for causal modeling of regulatory networks. Two approaches are described; one based on observational data, and other that combines observational and interventional data.

8.2 Learning a Causal Model for ADHD

Attention-deficit/hyperactivity disorder (ADHD) is a developmental disorder marked by persistent symptoms of inattention and/or hyperactivity and impulsivity that interfere with functioning or development. Symptoms begin in childhood and can affect daily life, including social relationships and school or work performance. Symptoms of ADHD can be mistaken for emotional or disciplinary problems or missed entirely in children who primarily have symptoms of inattention, leading to a delay in diagnosis. Researchers are not sure what causes ADHD, although many studies suggest that genes play a large role. Like many other disorders, ADHD probably results from a combination of factors.

In [9], the authors extended the BCCD algorithm for a mixture of discrete and continuous variables, and apply it to a data set that contains phenotypic information about children with Attention Deficit Hyperactivity Disorder (ADHD). They used a reduced data set that contains data from 223 subjects, with the following nine variables per subject:

1. Gender (male/female)
2. Attention deficit score (continuous)
3. Hyperactivity/impulsivity score (continuous)
4. Verbal IQ (continuous)
5. Performance IQ (continuous)
6. Full IQ (continuous)
7. Aggressive behavior (yes/no)
8. Medication status (naïve/not naïve)
9. Handedness (right/left)

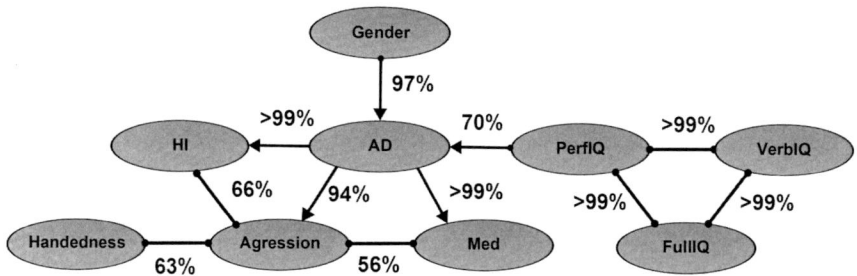

Fig. 8.1 The causal model obtained from the ADHD data set. The graph represents the resulting PAG in which edges are marked as → or − for invariant relations and as "o" for non-invariant relations. The reliability percentage of each edge is indicated. Notation: AD—attention deficit score, HI—hyperactivity/impulsivity score, PerfIQ—performance IQ, VerbIQ—verbal IQ, Med—medication status. Figure based on [9]

Given the small data set, they included some background information, in particular that no variable in the network can cause "gender". Using a reliability threshold of 50%, the network depicted in Fig. 8.1 was obtained; which is represented as a PAG (Partially Acyclic Graph).

Several interesting causal relations are suggested by the resulting network, some of which were already known from previous medical studies [9]:

- There is a strong effect from gender on the level of attention deficit.
- The level of attention deficit affects hyperactivity/impulsivity and aggressiveness.
- Handedness (left) is associated with aggressive behavior.
- The association between performance IQ, verbal IQ and full IQ is explained by a latent common cause.
- Only the performance IQ has a direct causal link with attention deficit.

8.3 Causal Discovery of Risk Factors for COVID-19

The COVID-19 pandemic disrupted the social and economic state of most countries worldwide. Data analysis can provide valuable information on the pandemic, including the most critical risk factors for hospitalization and morbidity; which could be useful for subsequent cases of COVID or other pandemics. In the case of Mexico, a large portion of its population is particularly vulnerable to the virus as a consequence of having diabetes, hypertension, and obesity; so in this work [10] the causal factors affecting the Mexican population were analyzed.

The Mexican National COVID-19 Data Base incorporates organized and standardized epidemiological and demographic information on the evolution of the COVID-19 pandemic in Mexico. The data was collected by the Respiratory Viral Epidemi-

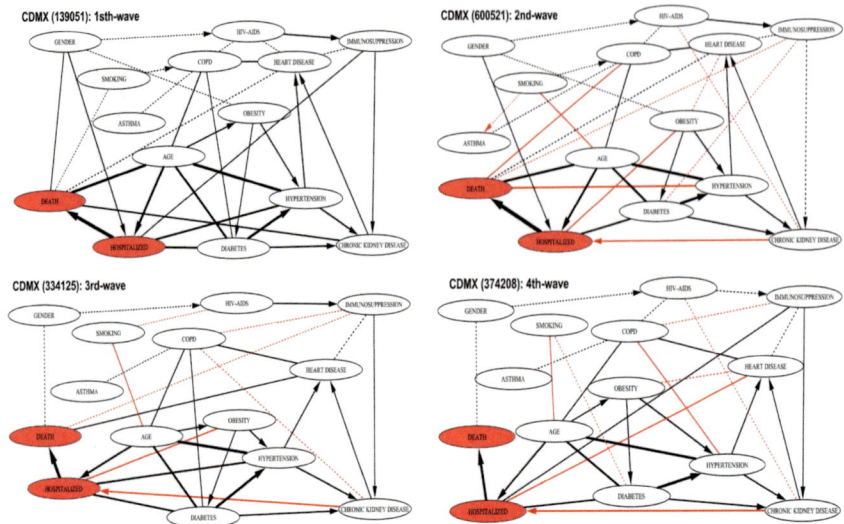

Fig. 8.2 Causal graphical models of Mexico City by wave. The number of cases used to learn each model is shown enclosed in parentheses. The thickness of the lines indicates the strength of the correlation between the variables; dotted lines represent a correlation factor of less than 0.05. Edges that are not included in the first wave model are shown in red

ological Surveillance System, consisting on 5,186 health units. It contains over 97 useful variables, including the general references of the patient, such as sex, age, place of residence; symptoms and comorbidities; and also data on testing, hospitalization, and deaths. The system includes more than 6.5 million individual records, available for research purposes at http://covid-19.iimas.unam.mx.

The fast causal inference algorithm (FCI) was used to uncover causal relation between several factors, in particular age, gender and comorbidities, as well as the risk of hospitalization and death. The analysis was performed for different waves of the pandemic, for two states in Mexico: Mexico City and Yucatán. Additionally, an analysis for different age groups was performed.

Figure 8.2 depicts the graphical causal models discovered for Mexico City for each of the four waves of the COVID pandemic. The graphical models obtained for Yucatán are shown in Fig. 8.3. Yucatán was a special case in Mexico where there were five waves. In the causal models, the strength of the relation is denoted by the width of the arcs, and the differences with the first wave are highlighted as red arcs. The main causal effects of interest, *HOSPITALAIZED* and *DEATH*, are shown as red ovals. We can notice that some arc remained undirected as the algorithm cannot determine the direction of causality or if there is an unmeasured cofactor that produces the correlation.

From the causal models we can verify some well known relations, such as if a patient is hospitalized the probability of death increases. Also, that age is an important factor related to the probability of hospitalization and death. Known relations between

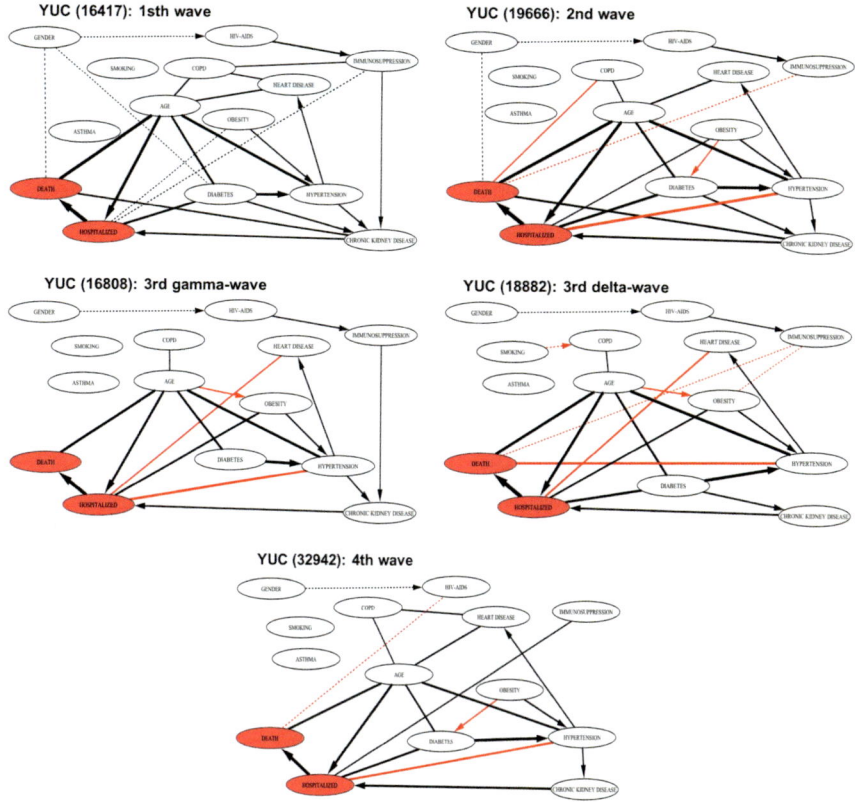

Fig. 8.3 Causal graphical models of Yucatán by wave. The number of cases used to learn each model is shown enclosed in parentheses. The thickness of the lines indicates the strength of the correlation between the variables; dotted lines represent a correlation factor of less than 0.05. Edges that are not included in the first wave model are shown in red

some of other variables were discovered, such as between obesity and diabetes, obesity and hypertension, and HIV and immunosuppression. Other not so obvious relations are discovered, such as:

- Chronic kidney disease seems to be a causal factor for hospitalization and death.
- The presence of diabetes is another factor for hospitalization and death, present in almost all the models.
- Hypertension seems correlated with hospitalization or death, although no causal direction was determined, this might be mediated by an unmeasured cofactor.
- Heart disease also appears correlated with hospitalization or death, in particular in the later waves of the pandemic.

We can also notice differences between the causal models across the different waves of the pandemic. This could be due to differences in the virus strains for the

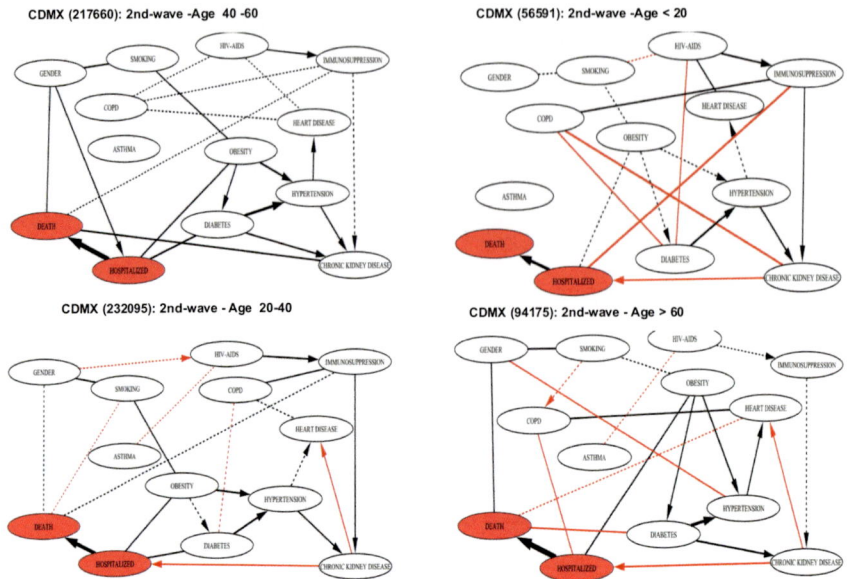

Fig. 8.4 Causal graphical models for different age groups in Mexico City during the 2nd wave of the pandemic. The number of cases used to learn each model is shown enclosed in parentheses. The thickness of the lines indicates the strength of the correlation between the variables; dotted lines represent a correlation factor of less than 0.05. Edges that are not included in the model of the age group between 40 and 60 years old are shown in red

different waves, and also the impact of the vaccines in the later stages. An interesting observation is that although gender had certain impact in the first waves of the disease, this tends to disappear in the latest stages.

Regarding the differences between Mexico City and Yucatán, we should clarify that Mexico City's models are more reliable, as the amount of data used to learn the causal models is much more (almost by an order of magnitude). Given the type of algorithm used (FCI), the causal graphs of Yucatán tend to be more sparse, although the main causal relations are similar in both states.

Figure 8.4 shows the results for different age groups in Mexico City. Some interesting observations:

- Gender seems to be a factor for the older age groups, but it disappears for the youngest age group.
- In the youngest group (age <20), the main factor for hospitalization were chronic kidney desease, immunosuppression and obesity.
- In the oldest group (age >60), the main factors that are responsible for increasing the probability of death are heart disease and diabetes.

8.4   Eliciting Brain Effective Connectivity

Neuronal physiological processes in various regions of the brain influence processes in other regions, this type of neural connectivity is known as *effective connectivity*. It can be seen as a causal process, as some regions *cause* the activation of other regions in the brain. Indirect measures of the local processes from neuro-images, result in time series whose statistical relations are often used as keys in attempts to recover causal relations among neuronal processes.

In this section, we describe two attempts to recover brain effective connectivity from neuro-images, one based on functional optical neuro-imaging and other that uses functional magnetic resonance imaging.

8.4.1   Decoding Brain Effective Connectivity Based on fNIRS

Montero et al. [5] explore the use of causal graphical models for decoding the effective connectivity of the brain from functional optical neuro-imaging (fNIRS). The basic idea is that directions of arcs in the connectivity network, that are left undecided (by existing learning algorithms) can be resolved by exploiting prior structural knowledge from the *human connectome*. A variant of the fast causal inference algorithm, seeded FCI, is therefore proposed to handle prior information.

To accomplish complex tasks, the functionally specialized brain regions have to collaborate with each other either by temporally coordinating their actions (functional connectivity), or causally regulating their activity (effective connectivity). Modeling the effective connectivity is important to understand the brain behavior and the expression of neural circuits. Effective connectivity is concerned with decoding cooperating brain regions, and most importantly, determining the direction of the flow of information; thus, it can be considered as a causal model, in which the activation of certain region in the brain, *causes* the activation of other region(s). The brain regions can be modeled as variables within a system; and their activity can be estimated based on fNIRS neuro-imaging, which is able to take a snapshot of the cortical activity across brain regions while a person is performing certain task.

In fNIRS, infrared light is irradiated at the scalp and travels through extra cerebral tissues to reach the cortex, and a fraction of it returns to surface due to backscattering. Changes in light attenuation are attributed to changes in concentration of hemoglobin, in turn, assumed to be representative of underlying neural activity.

The structural connections of the human brain establish a set of anatomical constraints with respect to the possible paths in its connectivity. This set of constraints can be obtained from the so called human connectome, which establishes the expected physical links in the human brain. So the human connectome provides prior knowledge for learning the causal structure related to the effective connectivity in the brain, by limiting the potential connections (links) in the causal network.

To incorporate prior knowledge, a modified version of the FCI algorithm was developed, named *seeded FCI* (sFCI). The addition of prior knowledge may resolve some of the undecided directions present in the output of the FCI algorithm, i.e., a

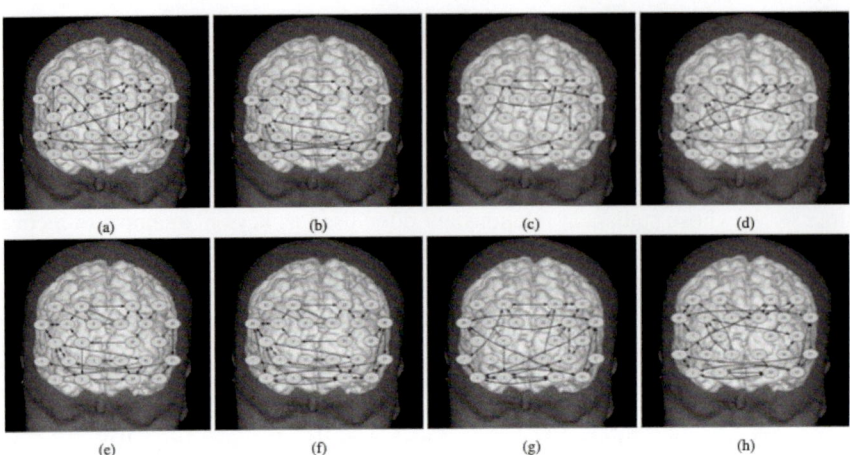

Fig. 8.5 Effective connectivity networks obtained for the knot-tying data set. The top row **a–d** depicts the results from the original FCI algorithm, while the bottom row **e–h** shows the results with sFCI which incorporated the prior knowledge from the connectome. Columns represent the different expertise groups: novices (**a**, **e**), trainees (**b**, **f**), consultants (**c**, **g**) and all (**d**, **h**)

Table 8.1 Number of undefined links in the learned PAGs for each group, using the FCI and sFCI algorithms

Algorithm	Novices	Trainees	Experts	All
FCI	11	19	18	26
sFCI	8	16	14	21

partial ancestral graph. This prior knowledge can be added in the form of restrictions by defining impossible relations. The basic idea in sFCI is to start with a complete undirected graph and a set of invariant links L (prior information). This additional information is incorporated in the original FCI algorithm and used as restrictions when removing and directing edges in the different phases.

The sFCI algorithm was applied to uncover effective connectivity in an fNIRS neuro-imaging data set collected at Imperial College to question about experience-dependent differences on pre-frontal activity for a group of surgeons while knot-tying. 62 surgeons (19 consultants, 21 trainees and 22 medical students) participated in the study, and brain activity was monitored with a 24-channel fNIRS system. A total of eight causal networks, shown in Fig. 8.5, were learned considering four groups (novices, trainees, experts and all subjects) and two variants of the FCI algorithm, with and without prior knowledge. The results are summarized in Table 8.1. As expected, the number of undefined links decreases with the utilization of prior information.

8.4.2 Estimating Effective Connections from fMRI Time Series

In Sanchez-Romero et al. [7], the authors compare different algorithms to recover effective connectivity in the brain from functional magnetic resonance imaging (fMRI) data. Initially, they applied the algorithms to simulated data, and then they evaluated two novel techniques, FASK and Two-Step, with two experimental data sets, one fMRI resting-state data set from the medial temporal lobe from 23 individuals, and one fMRI data set for a rhyming task from 9 individuals. Here we describe the results with the fMRI data, refer to the original paper for the simulation results.

8.4.2.1 Experimental Data

Resting-state fMRI data for 23 human subjects was acquired at temporal resolution, TR = 1 sec, for 7 min, resulting in 421 datapoints per subject. Seven regions of interest from the medial temporal lobe in each hemisphere were incorporated in the study: perirhinal cortex divided into Brodmann areas 36 and 35 (BA35 and BA36); parahippocampal cortex (PHC); entorhinal cortex (ERC); subiculum (SUB); cornu ammonis 1 (CA1); and a region comprising CA2, CA3 and dentate gyrus together (CA23DG). The FASK and Two-Step causal discovery algorithms were run on 23 repetitions of 10 individual subjects concatenated (4,210 datapoints) for the seven medial temporal lobe regions of interest of the left and right hemispheres, separately. For comparison, they ran both algorithms on the 23 subjects datasets individually (421 datapoints).

The task data included nine subjects which judged whether a pair of visual stimuli rhymed or not. In each 20 s block, eight pairs of words were presented for 2.5 s each. Four blocks of words were followed by four blocks of pseudo words. Data was acquired with a 3T scanner, with TR = 2 s, resulting in 160 datapoints. Eight regions of interest were considered: left and right occipital cortex (LOCC, ROCC); left and right anterior cingulate cortex (LACC, RACC); left and right inferior frontal gyrus (LIFG, RIFG); and left and right inferior parietal (LIPL, RIPL). They included an *Input* variable built by feeding the rhyming task boxcar model with a canonical hemodynamic response function. If the algorithms infer edge orientations correctly, then edges from the *Input* variable must feedforward into some of the regions of interest, and no edge should point backward into the *Input* variable. Given the small number of subjects and reduced sample size (160 datapoints), the FASK and Two-Step algorithms were trained on a joint data set made of data from the nine individual subjects(1,440 datapoints).

8.4.2.2 Algorithms

Two novel algorithms are presented that can discover cyclic structures: Fast Adjacency SKewness (FASK) and Two-Step [7].

Fask

FASK is based on the Fast Adjacency Search stable (FAS-stable) algorithm as an adjacency estimator. The FAS-stable algorithm is the order-independent adjacency search of the PC algorithm that avoids spurious connections between parents of variables. It builds an undirected graph by iteratively testing conditional independence relations under an increasing size of the conditioning set.

The idea of the FASK algorithm is as follows: first, FAS-stable is run on the data, producing an undirected graph, G. Then, each of the $X - Y$ adjacencies in G is oriented as a 2-cycle $X \leftrightarrow Y$, or $X \rightarrow Y$, or $X \leftarrow Y$. It tests to see whether the adjacency is a 2-cycle by testing if the difference between $corr(X, Y)$ and $corr(X, Y \mid X > 0)$, and $corr(X, Y)$ and $corr(X, Y \mid Y > 0)$, are both significantly different from zero. If so, the bidirected edge $X \leftrightarrow Y$ is added to the causal graph. If not, the Left-Right orientation rule is applied to orient $X \rightarrow Y$, otherwise $X \leftarrow Y$.

Two-Step

The Two-Step algorithm represents a linear causal model, with possible unmeasured cofounding variables and cyclic relationships:

$$\mathbf{x} = \mathbf{Bx} + \mathbf{Mc} + \mathbf{e} \qquad (8.1)$$

where $\mathbf{x}$ is a vector of observed variables; $\mathbf{B}$ is the connectivity matrix defining the causal structure between observed variables; $\mathbf{M}$ is a matrix indicating which observed variables are affected by unmeasured confounders; $\mathbf{c}$ is a vector of unmeasured confounders, if they exist; and $\mathbf{e}$ is the vector of mutually independent noise terms.

Two-Step applies the principles of independent component analysis (ICA) by representing the observed variables as structural equations that include a set of unobserved components (see Sect. 4.3.5). The term $(\mathbf{Mc} + \mathbf{e})$ defines the unobserved components. Generally, this set of equations does not follow the ICA model, since the components defined by $(\mathbf{Mc} + \mathbf{e})$ are not necessarily mutually independent because of the possible presence of unmeasured confounders. The components in these terms can be divided into mutually independent variables or groups of variables, where the variables within the same group are not independent. Each of these groups are a set of confounders plus noise terms that influence one another, or are influenced by some common unknown mechanism. Under mild assumptions, the solution can be found by applying ICA and then testing for the independence between the ICA outputs. Otherwise, if there are no confounders, the standard ICA model is applied.

The next section presents the results obtained with both algorithms on the fMRI data.

8.4.2.3 Results

Resting-state

The FASK and Two-Step algorithms were run on the fMRI resting state data described earlier. When evaluating with empirical data, there is not a ground truth to assess

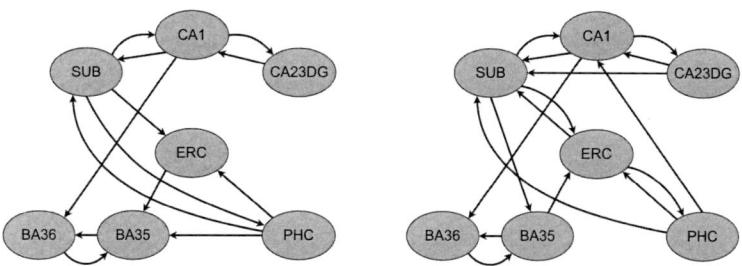

Fig. 8.6 Causal (effective connectivity) networks obtained by the FASK (left) and Two-Step (right) algorithms, showing only the robust edges. The brain regions of interest include cornu ammonis 1 (CA1); CA2, CA3, and dentate gyrus together (CA23DG); entorhinal cortex (ERC); perirhinal cortex divided in Brodmann areas (BA35 and BA36); and parahippocampal cortex (PHC). Figure based on [7]

the performance of the causal discovery algorithms. Instead, we have partial knowledge about the presence of structural connections between brain regions derived experimentally in animal models, and in *vivo* in humans by using diffusion imaging–based tractography [7]. This partial information about the connectivity can be used to evaluate (to certain degree) the performance of the causal discovery algorithms; for example, by examining the presence or absence of directed edges relative to the previous knowledge.

The causal discovery algorithms were applied to 23 repetitions of 10 individual subjects selected at random, standardized individually, and concatenated. The resulting 23 datasets include seven regions of interest from the medial temporal lobe, with 4,210 data points. Left and right hemispheres were analyzed separately. The graphical models obtained are combined by keeping only the *robust edges*. A robust edge is an edge estimated in 48% or more of the 23 repetitions of the 10 concatenated subjects. Figure 8.6 depicts the causal graphs for the seven regions of interest from the left hemisphere medial temporal lobe obtained with both algorithms, showing only the robust edges.

Overall, both algorithms produce a closely similar set of robust edges for the medial temporal lobe left hemisphere data. With one exception, every directed edge found by FASK is also found by Two-Step. The adjacencies and orientations are consistent with the medial temporal lobe model presented in previous works, capturing the flow of information from the medial temporal lobe cortices (BA35, BA36, PHC) directly into the entorhinal cortex (ERC), which works as a gateway to the hippocampal formation, where the signals travel to CA23DG to CA1 to the subiculum (SUB) and back to ERC, and from ERC back to BA35, BA36, and PHC. The presence of 2-cycles in the output of both algorithms is consistent with reported feedback in the medial temporal lobe. There are two main discrepancies of the results with the standard model of the hippocampus [3]:

1. Neither FASK nor Two-Step robustly inferred the ERC–CA23DG edge (perforant pathway). This is unexpected since this is the main pathway connecting the medial

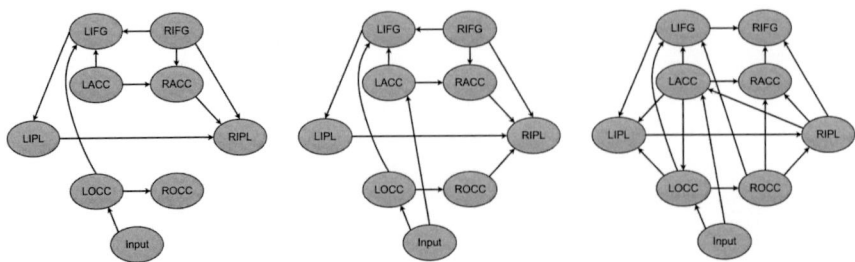

Fig. 8.7 Causal (effective connectivity) networks obtained by the FASK-FAS-stable (left), FASK-FAS-original (center), and Two-Step (right) algorithms, in the task experiment for eight bilateral regions of interest and one input variable. Regions of interest include left and right occipital cortex (LOCC, ROCC); left and right anterior cingulate cortex (LACC, RACC); left and right inferior frontal gyrus (LIFG, RIFG); and left and right inferior parietal (LIPL, RIPL). Figure based on [7]

temporal lobe cortices in the hippocampus. A possible explanation is that the signal between these regions measured with resting-state fMRI is not strong enough to be captured by the algorithms.

2. The second discrepancy is related to the CA23DG–SUB edge which shows in the Two-Step output but not in the FASK causal graph.

Although there are some differences, the causal discovery algorithms obtain causal models that are in general consistent with what is known from previous experimental studies, based on a relatively small data set.

Task Data

The FASK and Two-Step algorithms were run on one repetition of 9 subjects concatenated (1,440 data points) from the rhyming task data set, for the eight bilateral brain regions and one *Input* variable, representing the dynamics of the stimuli presentation. Two versions of FASK were executed, using FAS-stable and FAS-original (the order dependent adjacency search of the PC algorithm). The resulting causal graph for each algorithm is depicted in Fig. 8.7.

Both, FASK and Two-Step algorithms correctly output feedforward edges from the *Input* variable to the regions of interest (an important test to verify that the algorithms produce a correct orientation of the edges). Of particular importance is the edge going from the *Input* to the left occipital cortex, which is expected as this is a task in which the visual stimuli are initially presented to the subjects. Both algorithms present feed-forward edges from the left occipital cortex to the frontal lobe (LIFG, left inferior frontal gyrus), and then back to the parietal lobe (LIPL, left inferior parietal). This causal order is consistent with the rhyming task. The graph produced by the Two-Step algorithm is more dense, and the adjacencies of FASK (FAS-stable) and FASK (FAS-original) are proper subsets of the Two-Step adjacencies. In general, the orientations of the edges in these learned models are consistent with what is expected, in particular regarding the flow of information from the occipital to the frontal and to the parietal brain regions.

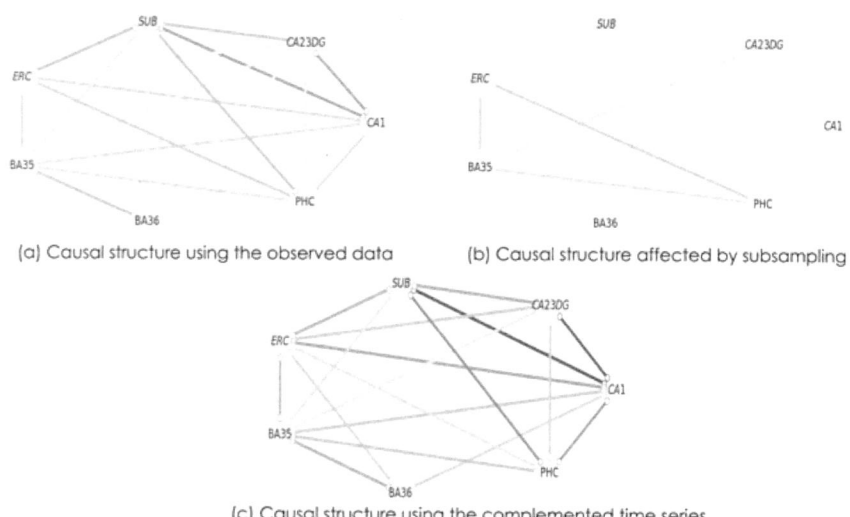

(a) Causal structure using the observed data   (b) Causal structure affected by subsampling

(c) Causal structure using the complemented time series

Fig. 8.8 Causal structures learned from observed data from the resting state of the left hemisphere medial temporal lobe: **a** original causal structure, **b** causal structure obtained with the sub-sampled time series, and **c** causal structure learned with the imputed data

8.4.2.4 Analysis Considering Subsampling

A relevant problem when using neuroimaging data for causal discovery, is the problem of subsampling (see Chap. 6), as some imaging modalities have a low sampling rate, so the acquisition rates tend to be slower that the interaction processes in the brain. This is a limitation specially in fRMI, as the sampling frecuency is in the order of one second.

An additional experiment using the same data for the resting state condition described above, considers the problem of subsampling. This data comprises the interaction of seven regions from the left hemisphere medial temporal lobe during resting state. The data set contains 4,120 datapoints for each brain region, which are used to generate the original causal structure. Subsequently, the time series is affected to simulate the effects of a subsampling rate of 2. Following the approach described in Chap. 6, the time series is complemented to obtain a time series with imputed data that allows to recover the causal structure. The PCMCI Algorithm (see Sect. 6.3.2.1) is applied to the subsampled and the complemented time series, the resulting structures are depicted in Fig. 8.8.

The adjacency matrix was used to compare the resulting causal structures. The original causal structure was compared to the causal structure affected by subsampling, resulting in Mean Average Error (MAE) of 0.3265. The causal structure of the complemented time series was also compared, resulting in an MAE of 0.0204; a significant improvement.

8.5 Regulatory Networks

Gene regulatory networks (GRNs) characterize the regulatory mechanism of genes in the form of a network in a cell. These networks provide an essential clue for the understanding of cell procedures and the molecular interplay that governs cell phenotypes. Gene regulatory network (GRN) inference from gene expression data is a significant approach to understand aspects of the biological system. We can think of this type of networks as causal networks, since certain genes *cause* the activation or deactivation of other genes. Causal discovery of regulatory networks is a relevant problem, as directed causal edges in a network allow us to identify pathways, or mechanisms of action, from genotype to phenotype. This is a challenging problem for causal discovery algorithms, given the large number of genes (variables) in even the simplest organisms, in the order of 1,000 to 10,000 genes. Additionally, the gene expression data used to learn them is usually quite limited.

Next, we describe two applications of causal discovery to infer gen regulatory networks, one based on only observational data and other that combines observational and interventional data.

8.5.1 Gene Regulatory Network Inference Based on Causal Discovery

Feng et al. [1] proposed a method for learning GRNs based on pairwise causal discovery, that combines gene expression data and gene representation derived from a graph neural network (GNN) model.

The proposed framework includes three phases: (i) a potential network is constructed from gene expression data, where an edge represents a strong correlation between genes, (ii) they sample positive and negative edges from the potential network and then train GraphSAGE (an improved graph neural network) to obtain the new representation of each gene by using binary classification as the downstream task, (iii) genes' representation from the well–trained GraphSAGE is fed into a modified additive noise model (ANM) to infer causal regulatory relationships.

The method was tested on 15 datasets, including three benchmark datasets for which the gene regulatory network is known and can be considered as gold standards. These are: (i) *In silico*, which is a synthetic dataset whose expression matrix is generated by the simulation software, (ii) *E.coli*, that provides experimentally validated interactions from the curated database RegulonDB, and (iii) *S.cere*, that includes a set of interactions supported by genome–wide binding data. Table 8.2 summarizes the main properties of the three datasets.

They evaluated the proposed method by a comparison with the gold standard based on the *area under the receiver operating curve* (AUROC), obtaining the following results: (i) *In silico*, 0.725, (ii) *E.coli*, 0.626, and (iii) *S.cere*, 0.571. These results are superior to other approaches, in particular for the E.coli and S.cere datasets (see the original paper [1] for more details). Still, the results are not very good given the

Table 8.2 Properties of the three datasets used to validate the method

Dataset	#Samples	#Genes	#Edges	Avg. out-degre	Avg. in-degree
In silico	805	1643	4012	20.57	2.44
E.coli	805	4511	2066	6.19	0.46
S.cere	536	5950	3940	11.83	0.66

challenge of learning a causal network with a high number of variables and limited data, as we mentioned before. An alternative is to use interventional data.

8.5.2 Learning Biological Networks from Observational and Interventional Datasets

Shah et al. [8] proposed a hybrid structure learning method that jointly learns from observational and interventional data. By jointly learning from observational and interventional data, they can correctly isolate causal relationships and orient edges in the graph. Given that the space of possible interventions is very large, they also consider an optimal experimental design to choose the interventions that can be more useful to drive causal discovery.

The method's goal is to recover the underlying Causal Bayesian Network given a mix of interventional and observational samples, and some prior information about the causal edges in the graph. They represent this problem as a Bayesian Inference problem over the space of DAGs, incorporating as prior knowledge, $P(G)$, any prior structural knowledge about the causal graph. This is combined with the observational and experimental data, D, via Bayes theorem: $P(G \mid D) \propto P(G)P(D \mid G)$.

They incorporate *hard* interventions (also known as structural or ideal interventions, see Sect. 5.2), where a node is set to a constant value, because this mimics the types of interventions performed, e.g., gene knockout experiments.

The method consists of two steps:

1. A causal discovery algorithm (PC, ARACNE, CLR) is applied to the observational samples to generate the skeleton of the causal graph.
2. Based on the skeleton obtained from (1), Greedy Interventional Equivalence Search (GIES), which is a greedy score-based approach that is an extension of the original GES learner, is applied to jointly learn from the observational and interventional data. Other algorithms were also considered.

The method was evaluated on the DREAM4 insilico challenge that provides observational and interventional gene knockout data of gene regulatory networks synthetically generated. The data set contains 10 observational samples and one interventional sample per node. The ground truth networks are subnetworks of the larger E.coli gene regulatory network. It was also tested on larger scale networks, based

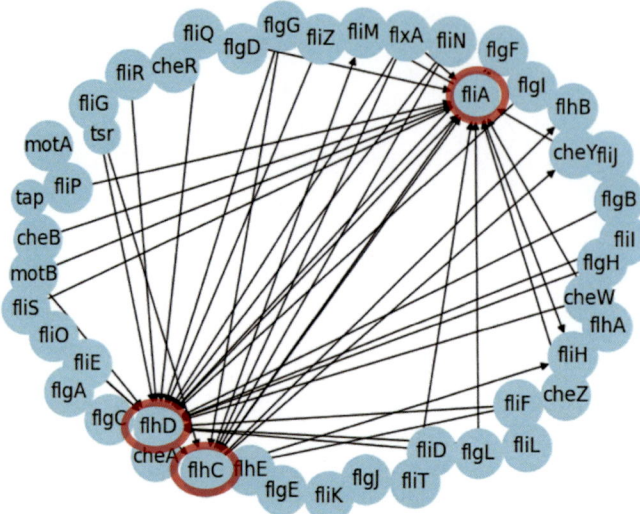

Fig. 8.9 A subnetwork estimated by SP-GIES using the CLR skeleton from the RegulonDB data extended with synthetic data. Three hubs are detected (in red), although the graph is much more sparse than the ground truth network. Figure from [8]

on the RegulonDB dataset. The network is the current estimated E.coli K12 gene regulatory network and contains 1146 nodes, 3179 edges and 524 real experimental samples. Of the 524 experimental samples, 35 are samples are taken from gene knockout experiments.

They tested several combinations of learning algorithms, evaluated with three metrics: Structural Hamming Distance (SHD), Structural Interventional Distance (SID) and the Area Under the Precision-Recall Curve (AUC-PR). The algorithms based on observational data that were evaluated are: PC, ARACNE, and CLR; and the algorithms based on interventional data that were evaluated are: GIES, IGSP, Pinna and SP-GIES. The results vary depending on the size and the type of networks. For DREAM4 insilico size 10, they achieved an AUC-PR over 0.6; for synthetic data with 1000 nodes and 1000 edges it reached 0.91 AUC-PR in one case, and for RegulonDB 1146 nodes, 3179 edges, the best result was 0.5 AUC-PR. One of the causal networks obtained is depicted in Fig. 8.9, which corresponds to a subnetwork of the E.coli.gene regulatory network learned from the RegulonDB data set.

One of the main challenges is computational complexity. For instance, running a network of 10,000 nodes, GIES could not finish in 24 h. It seems that the type of structure of the network has a greater impact than the number of variables. The performance on the RegulonDB data set, which is the most biologically relevant data set, suggests that future work should also incorporate nonlinearities.

8.6 Additional Readings

Glymour et al. [2] present a review of causal discovery methods based on graphical models and several applications to biology. Ma et al. [4] discuss several topics related to causal discovery from biomedical data, including causal structural learning from observational and experimental data. In [6] four algorithms for genotype–phenotype network structure learning are discussed.

Acknowledgements This chapter is based on the work of different groups as referenced in the corresponding sections. Figure 8.9 used with permission.

References

1. Feng K, Jiang H, Yin C, Sun H (2023) Gene regulatory network inference based on causal discovery integrating with graph neural network. Quant Biol 11:434–450
2. Glymour C, Zhang K, Spirtes P (2019) Review of causal discovery methods based on graphical models. Front Genetics 10, Article 524
3. Lavenex P, Amaral DG (2000) Hippocampal-neocortical interaction: a hierarchy of associativity. Hippocampus 10(4):420–430
4. Ma S, Statnikov A (2017) Methods for computational causal discovery in biomedicine. Behaviormetrika 44:165–191
5. Montero-Hernandez SA, Orihuela-Espina F, Herrera-Vega J, Sucar LE (2016) Causal probabilistic graphical models for decoding effective connectivity in functional near infrared spectroscopy. In: The twenty-ninth international FLAIRS conference. AAAI Press
6. Ribeiro AH, Soler JMP, Neto EC, Fujita A (2016) Causal inference and structure learning of genotype-phenotype networks using genetic variation. In: Wong KC (ed) Big data analytics in genomics. Springer, Cham
7. Sanchez-Romero R, Ramsey JD, Zhang K, Glymour MRK, Huang B, Glymour C (2019) Estimating feedforward and feedback effective connections from fMRI time series: assessments of statistical methods. Netw Neurosci 3(2):274–306. The MIT Press
8. Shah A, Ramanathan A, Hayot-Sasson V, Stevens R (2023) Causal discovery and optimal experimental design for genome-scale biological network recovery. arXiv:2304.03210v1
9. Sokolova E, Groot P, Classen T, Heskes T (2014) Causal discovery from databases with discrete and continuous variables. In: Probabilistic graphical models (PGM). Springer, pp 442–457
10. Sucar LE, Serrano-Pérez J, Rodríguez-López V, Gutierrez-Rios R, Pineda LA (2021) Prediction and analysis of COVID-19 with probabilistic graphical models. In: Bayesian modeling application workshop, 37th conference on uncertainty in artificial intelligence

Applications in Social Sciences

9

Abstract

This chapter reviews several applications of causal discovery in social sciences. One application deals with understanding the factors that people take into consideration when making decisions related to the energy transition. A second example presents a causal model relating environmental factors to armed conflicts. A third example deals with discovering causal relations in the stock markets. Other application is about finding causal relations from legal texts, and finally, learning causal structures for marketing mix models.

9.1 Introduction

Social science studies how societies work, how people interact with one another. The branches of social science include anthropology, economics, political science, psychology, sociology, linguistics, education, history, and law. In some of these domains, causal models can have an important impact as we can predict the results of interventions using a causal model, before the interventions are done in *reality*. This could imply an important saving in time and resources.

© The Author(s), under exclusive license to Springer Nature Switzerland AG 2026   187
L. E. Sucar, *Causal Discovery*, Computer Science Foundations and Applied Logic,
https://doi.org/10.1007/978-3-031-98345-0_9

In this chapter we analyze several examples of the application of casual discovery in different branches of social sciences:

- Developing a causal model including the different factors that affect households' decisions for the energy transition. In this case, the causal model is constructed based on expert knowledge.
- Understanding the causal mechanisms that produce environmental conflicts, combining literature review with causal discovery.
- Discovering causal relations between stock prices in the stock markets, and using these to propose a trading strategy.
- Finding causal relations from legal texts, and applying them to similar charge disambiguation.
- Learning causal structures for marketing mix models.

These applications illustrate different strategies for the acquisition of causal knowledge in several contexts; including: (i) building a causal model from experts, (ii) combining literature and observational data, (ii) causal discovery from temporal data, (iv) discovering causal relations from unstructured data, i.e., text, and (v) causal discovery through graph neural networks.

9.2   Understanding the Factors that Affect Households' Decisions for the Energy Transition

Together with greener energy sources, reducing energy consumption is an important aspect for controlling emissions and global warming. The importance of human behavior has frequently been seen as a minor or even unimportant consideration. Although the integration of the human component is still in its infancy, it has a great potential for encouraging the adoption of greener behaviors at home.

Aguayo-Mendoza et al. [1] introduce a methodology that retrieves human expert knowledge from four key aspects of the energy transition[1]: everyday appliances, buildings, mobility, and energy efficiency. Given the lack of data regarding the factors that people take into consideration for the energy transition, the proposed methodology, based on the Delphi Method, involves conducting a qualitative study to create a taxonomy of the most relevant factors that affect households' investment decisions, regarding the energy transition. It aims to build a causal model that identifies the main factors that impact families and individuals when making energy investments related to any aspect of the energy transition.

---

[1] The energy transition refers to the shift from fossil fuel-based energy systems to more sustainable, low-carbon energy sources, like renewable energy (solar, wind, hydro, and geothermal) and cleaner technologies. The transition is critical for reducing climate change impacts, improving energy security, and creating a more sustainable global economy.

A taxonomy of factors that influence energy-related decision-making was obtained. A cross-sectional survey was then conducted to determine the characteristics of the European population regarding behavioral factors related to energy transition. The study aimed to identify clusters of individuals with similar determinants that motivate their energy-related decisions, which could be targeted by policy measures. It examined the relationship between energy consumption behavior and socio-demographic variables like gender, age, education level, and household income.

To create the causal diagram, they focus on four use cases or application domains: appliances, building renovation, flexibility services, and mobility. Then they developed five hypothetical scenarios for each field. In these scenarios, each *expert* must identify factors that influence the people's investment decisions. The scenarios are meant to represent different realities:

1. Baseline: These scenarios aim to depict a typical European city house in its current state.
2. Minimum: Refers to the least amount of effort required to make progress toward the decarbonisation of specific areas of application, including appliances, building renovations, flexibility services, and mobility.
3. Probable: These scenarios consider the most probable decarbonisation actions that citizens in Europe would take over the next few years.
4. Plausible: These scenarios are less probable but not too unusual for some European households working towards decarbonisation goals.
5. Ideal: It is highly unlikely that these ideal scenarios will be achieved because significant social and cultural changes are required.

After validating the scenarios, they created a comprehensive list of relevant factors that influence households' decisions to invest time and money to make these scenarios a reality. A group of experts identified 32 determinants (factors) that provide compelling reasons for individuals to invest their time or money in each of the given scenarios. Nine main factors were identified with several associated factors:

- Autonomy: adherence, self-satisfaction, commitment.
- Relatedness: support, agreement, socializing, peer pressure, right and duties.
- Physicalness: wellbeing, cozyness.
- Security; trust, safety, legal, climate protection.
- Meaning: own significance, environmental concern.
- Popularity: poseur, trends.
- Financial: risk profile, added value, profits, credit score, frugality.
- Competence: cost efficiency, knowledge, own competence, technical fit.
- Simulation: brag, fun, novelty.

According to experts, Competence factors and Relatedness are the most significant factors, surpassing financial, and a survey confirms this view. Although experts consider Relatedness as the second-order priority, survey respondents prioritize Autonomy and Security as second and third, respectively, ahead of financial needs. It seems that internal factors such as beliefs, values, or environmental concerns are the prominent drivers that foster a change in people's daily behavior to invest time and effort in reducing carbon and energy footprint.

Based on the researchers' expertise and the 32-factor taxonomy, a set of initial *archetypes* was proposed [2]. Archetypes are universal, inborn models of people, behaviors, and personalities that play a role in influencing human behavior. A total of eight archetypes were defined for the European population[2]:

1. The early adopter. An archetype who is driven by novelty.
2. The uninterested. The maintenance of comfort is what predominates its daily life when making decisions.
3. The *Homo Economicus*. Its motivations for undertaking any activity are mainly economic, either to make profits or to reduce expenses.
4. The fearful. Is usually afraid of making decisions, or lazy, which entail inaction, because of the amount of risk and time involved.
5. The stubborn. A person who is highly committed to environmental issues. However, this situation makes her feel distressed, anxious, or angry.
6. The influencer. Enjoys influencing peers. It looks always for an added value to her actions, either monetary, authoritative, or increased social capital.
7. The careful. A person that tries to maximize personal, collective, and ecosystem well-being and security.
8. The activist. Has the compromise of supporting his community. Has pro-environmental values, and beliefs that are aligned with its way of living.

For each archetype, the determining factors that affect their decisions were identified.

To map the most relevant factors for each archetype, the *Transtheoretical Model of behavioral change* (TTM) was used. TTM characterizes behavioral change as a process that unfolds over time, a non-linear transition through a sequence of qualitatively distinct stages: (i) precontemplation, (ii) contemplation, (iii) preparation, (iv) action, (v) maintenance, (vi) completion. Then, the most relevant determinants for each of the different phases of the TTM on each initial archetype were identified. By joining the main factors for each stage in a directed graph, a causal model for each archetype was constructed. The causal graphs for some archetypes are depicted in Figs. 9.1, 9.2, and 9.3 (the causal graphs for all the archetypes can be found in [2]).

Once these models are validated and parameterized, they could be used to define an intervention strategy to promote energy transition in households, by evaluating the different possible actions and their impact on the behavior of people in different regions.

---

[2] Similar archetypes were obtained for the Latinamerican population.

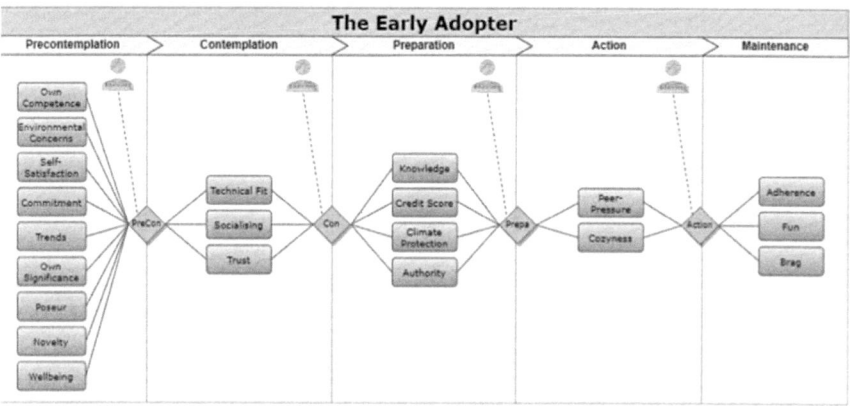

Fig. 9.1 Causal graph for the *early adopter* archetype. Figure taken from [2]

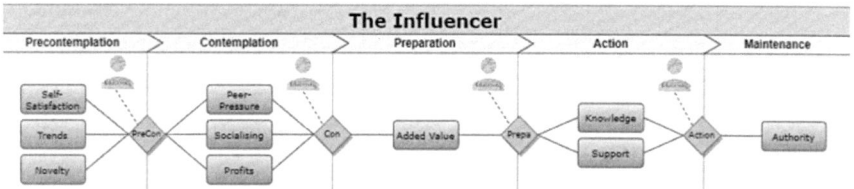

Fig. 9.2 Causal graph for the *influencer* archetype. Figure taken from [2]

Fig. 9.3 Causal graph for the *activist* archetype. Figure taken from [2]

9.3 Causal Discovery in Environmental Conflict

Vonk [13] presents a causal discovery application to environmental conflict, in particular for the conflict dynamics in Iraq, where an extensive literature review is paired up with a causal discovery algorithm to uncover causal mechanisms.

Environmental conflict is a form of tension or violence triggered by competition over natural resources or environmental pressures, such as scarcity or degradation. Although literature has advanced in examining the causal relationships between climatological factors and the onset of armed conflict, these studies often focus on specific connections rather than a comprehensive mechanism of the naturally induced conflicts. One way to address this gap is using causal discovery.

Based on a literature review [13], an hypothetical causal causal graphical model that represents the relations between environment and conflict was proposed, see Fig. 9.4. The model combines several factors, including: (i) long-term weather patterns that have been directly linked to armed conflict, (ii) effects of environmental changes on armed conflict mediated by the scarcity of vital resources, (iii) emergence or existence of armed conflict in relation to environmental changes and population sizes.

In addition to the hypothetical causal structure based on the literature, a causal model was learned from data related to the conflicts in Irak. The time horizon of the data was from the years 2020 and 2021, and includes the following variables, which were selected based on the previous literature review:

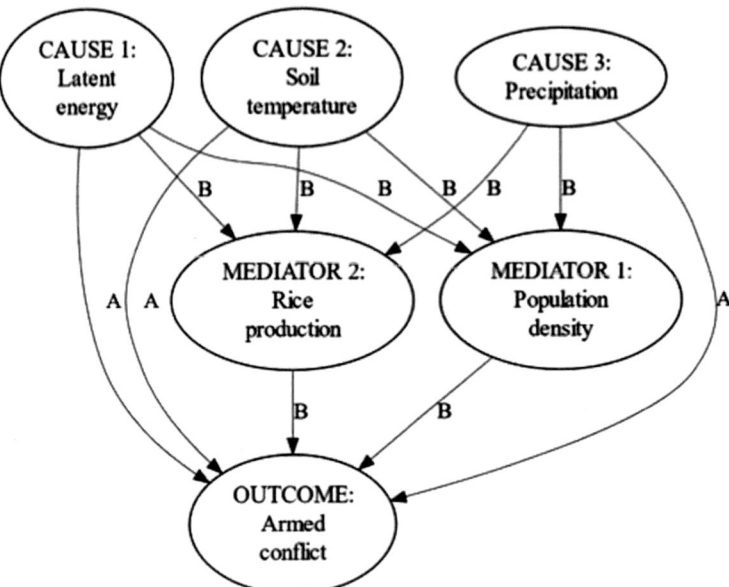

Fig. 9.4 Causal structure derived from the literature review. It distinguishes between direct (**a**) and indirect (**b**) links. Taken from [13]

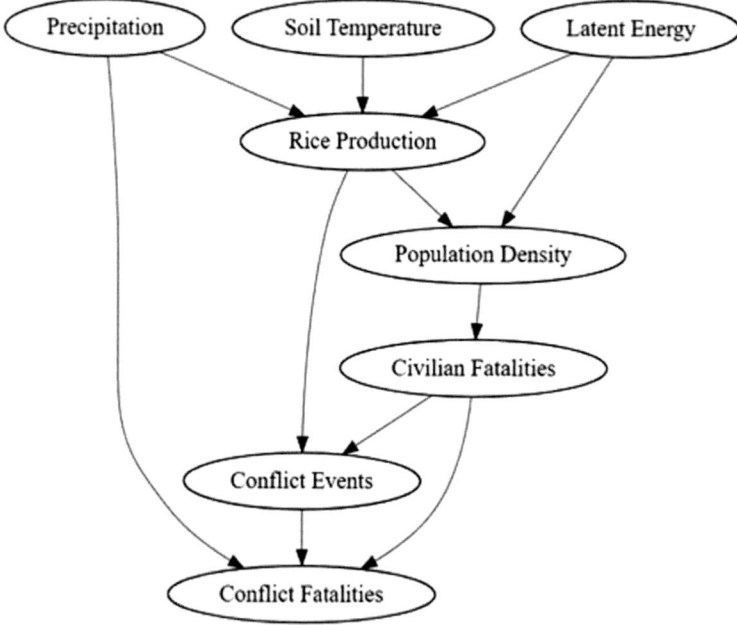

Fig. 9.5 Causal structure obtained by causal discovery based on the data from Irak using the GES algorithm. Taken from [13]

- Number of civilian fatalities, conflict events, and conflict fatalities, obtained from Armed Conflict Location and Event Data Program.
- Climatological-related variables such as precipitation, soil temperature, and latent energy.
- Rice production data from MapSPAM[3]
- Population density data from the International Earth Science Information Network at Columbia University.

Using the previous data, the GES algorithm was applied for causal discovery. GES was chosen as it has been shown to produce reasonable results in the case of simulation studies with small sample sizes. Several causal graphs were obtained in the causal discovery phase, and the one that shares the most directed edges with the causal graph based on the literature review was selected. This causal graphical model is depicted in Fig. 9.5.

The model obtained through causal discovery, reveals some more specific causal relations. It shows that precipitation is the only environmental variable with a direct effect on conflict, while other effects are mediated by rice production or population density.

---

[3] Spatial Production Allocation Model: https://mapspam.info/.

9.4 Predicting the Stock Market

Tang [12] analyses causality-based trading strategies in the stock market, and evaluates their feasibility and effectiveness. Previous research has applied time series
causal discovery techniques to uncover driving forces in the stock market; this work
goes beyond just identifying the causal factors, by predicting future stock prices
based on these driving forces. With these predictions, it becomes possible to manage
an investment portfolio with a causal-based trading strategy.

The empirical study included three aspects: (i) data collection and processing; (ii)
causal discovery, predictions and trading strategy; and (iii) evaluation.

The data compromises three stock markets, their characteristics are summarized
in Table 9.1. It includes China (CSI300) and the United States (SP500) as two major
stock markets, and the stock portfolio of Nancy Pelosi, who is well known for her
astute investment insights. Data imputation was performed due to the substantial
amount of missing values in the datasets. The imputation process involved two steps.
First, linear interpolation was applied across all stocks to estimate and fill gaps where
missing values occurred between known values. Second, any stock that still contained
missing values after interpolation was removed from the data set.

Three causal discovery algorithms, tsFCI, VarLiNGAM and TiMINo, were
employed to learn the causal structure from each data set. From the causal graphs,
the set of parent nodes of each stock X, denoted by $Pa(X)$, are obtained, which can
be interpreted as the driving forces of X. Then, a predictive model for each stock is
defined, using $Pa(X)$ as independent variables. The model is formulated with the
following equation for each stock:

$$P_t(X) = f_X(Pa(X_{t-1}), ..., Pa(X_{t-\tau})) \tag{9.1}$$

where τ is the maximum time lag used during causal discovery. Thus, the current
value of a stock is predicted by the changes in the relevant stocks at the previous
time steps.

For each stock, the historical values of its driving forces are obtained, then a linear
regression model on these past prices is fit, and then a one-step ahead prediction is

Table 9.1 Characteristics of the stock markets used in the study

Data set	Pelosi	CSI300	SP500
Market	US	China	US
No. Stocks	12	98	446
Start date	30-Jul-2019	1-Sep-2009	1-Sep-2009
End date	30-Jul-2024	31-Dec-2019	31-Dec-2019
No. days for training	1007	2010	2083
No. days for testing	252	503	521

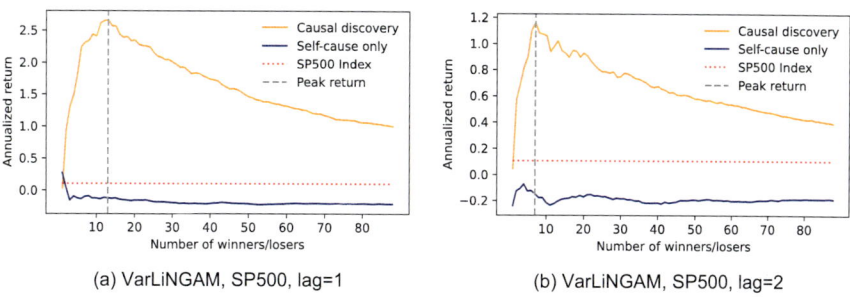

Fig. 9.6 SP500 portafolio return using VarLiNGAM. The graphs show the annualized returns against the number of winner/loser stocks considered in the trading strategy, for a time lag of 1 (**a**), and a time lag of 2 (**b**). Taken from [12]

performed. Using the predicted prices, they calculate the predicted returns for the next day as follows:

$$\gamma_{t,t+1}^{X} = \frac{p_{t+1}^{X} - p_{t}^{X}}{p_{t}^{X}} \tag{9.2}$$

where $\gamma_{t,t+1}^{X}$ is the predicted return, p_{t+1}^{X} is the predicted price of stock X for the next day, and p_{t}^{X} is the actual stock price today. The predictions of all stocks in the asset class for the next day are aggregated and the corresponding trading actions are selected.

Based on the predictions, the following trading strategy was implemented. First, they select n winner stocks, and also n loser stocks. Second, using the one-day-ahead predictions obtained through regression, they identify the n stocks with the highest and with the lowest predicted returns, classifying them as winners or losers. Finally, at the end of the current trading day, buy the winners and sell the losers.

Historical data was used to evaluate the performance of the trading strategy. To avoid look-ahead bias, the earliest 80% of the dataset is used as the training set, and the remaining 20% is used as the test set. In each trading day, they run the trading strategy and compute the outcome according to the real data. They continue with one-step ahead forecasts until the end of test period. Finally, the annualized portfolio return is calculated.

To simplify the comparison, all returns were standardized to annualized returns. Each asset class was benchmarked against its respective index to evaluate portfolio performance. As a baseline, a control portfolio was constructed using predictions based only on the previous values of the same stock.

They performed the tests with the three algorithms; however, given the computational resources required, only VarLiNGAM could finish within 24 h. Here we present the results for SP500 with VarLiNGAM, Fig. 9.6 (for the other see [12]). In this case, trading using the predictions based on the causal models are clearly superior to the SP500 Index and self-cause only, and the best results are with $n = 12$ for a time lag of 1, and $n = 8$ for a time lag of 2.

In summary, the results indicate that incorporating a causal discovery phase generally improves portfolio returns, especially with larger data sets, where profitability increases with data size. Among the three algorithms, VarLiNGAM is the most effective, handling large data sets and generating the highest returns on all data sets.

9.5   Leveraging Causal Inference in Legal Text Analysis

Most research in causal inference is devoted to analyzing structured data, and there is less work on the challenging area of discovering causal relationship from unstructured and high-dimensional data, like text. This is the case for applying causal discovery in legal text, as illustrated in Fig. 9.7.

In [9], a Graph-based Causal Inference (GCI) framework for legal documents analysis is proposed. This method builds causal graphs from fact descriptions without much human involvement, enabling causal inference to facilitate legal practitioners to make better decisions. GCI first recognizes key factors by extracting keywords from the text descriptions and it clusters similar ones as variables to be included in the causal graph. Then, it constructs a causal graph on these variables (nodes) based on a causal discovery algorithm that can tolerate unobserved variables. It then estimates the causal strength of each edge, and uses the refined graph to help decision making. The main focus of GCI is *similar charge disambiguation*, because an important challenge in judicial practice is to distinguish between similar charges.

GCI consists of four phases: (i) extracting the factors (variables) from the text, (ii) constructing the causal graph, (iii) estimating causal strengths, and (iv) making decisions.

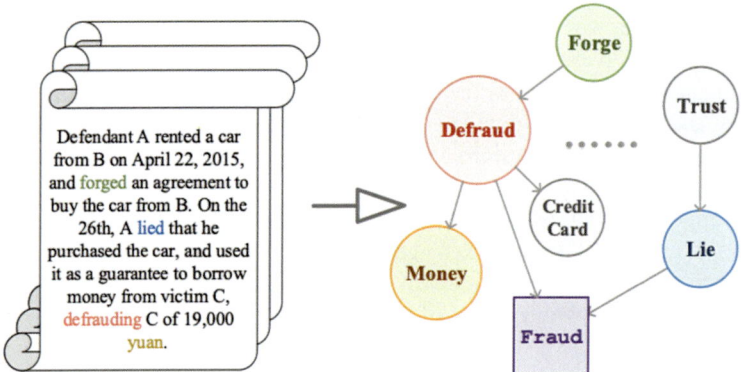

Fig. 9.7 An example of causal discovery from a legal text. Colored words in the text are matched between the exemplified fact description (a fraud case) and the nodes in the causal graph. Taken from [9]

Given the fact descriptions of criminal cases $D = \{d_1, d_2, ..., d_N\}$, the system classifies each case into one charge from the similar charge set $C = \{c_1, c_2, ..., c_M\}$. To extract the variables from the texts, it calculates the importance of word w_j for charge c_i using YAKE [5], enhanced with inverse document frequency to extract the more discriminative words of each charge. To discriminate the similar charges, it selects p words with the highest importance scores for each charge, clustering them into q classes to merge similar keywords. The q classes together with the M charges constitute the nodes of the causal graph. All the factors are binary variables.

For causal discovery, GCI incorporates the Greedy Fast Causal Inference (GFCI) algorithm [10]. GFCI is a hybrid method that combines score-based and constraint-based algorithms, and tolerates hidden confounders, generating a PAG. Given the presence of confounders, GCI uses the *Average Treatment Effect* (ATE) as a measure of causal strength.

Once the relevant variables are identified in the cases, the causal graphs are built. For this, GCI considers several heuristics. First, as the judgement is made based on the fact description, it does not allow edges from charge nodes to other ones, e.g., an edge from *fraud* to *lie* is prohibited. Second, given that causes usually appear before effects in time, and fact descriptions in legal text are often written in the temporal order of events, it considers the chronological order of descriptions as temporal constraints, providing a causal order that restricts edges according to this order.

Given that the output of GFCI is a PAG, there are several types of edges which include uncertain relations (see Sect. 4.2). To estimate the causal strength of each edge, GCI samples Q causal graphs from a PAG. For example, for an edge of type *either X causes Y, or an unobserved confounder*, it keeps an edge with 0.5 probability, and removes it with 0.5 probability in the generated graphs. The quality of each generated graph, G_q, is measured by its fitness with the data, using the Bayesian information criterion (BIC). Based on the set of generated graphs, the parameters are determined qualitatively based on the estimation of the ATE. It assigns high strength to edges with strong causal effect, and near-zero strength to edges that do not indicate causal relations or with a weak effect.

Finally, the causal graphs are used to make decisions regarding which charge, in the set of possible charges, C, is more appropriate for a case. For this, it extracts factors from the case in question, and maps the case description with the causal graph. First, it computes the overall causal strength of each factor to Y_i among the Q sampled causal graphs, where Y_i represents whether charge c_i is committed. Then, for each case, it maps the text with the graphs, and calculate scores for each charge. Finally, the calculated scores are fed into a random forest classifier to learn thresholds between the charges.

To evaluate the proposed framework, they selected five similar charge sets from the Criminal Law of the People's Republic of China, which are hard to discriminate in practice, and selected the corresponding fact descriptions from the Chinese AI and Law Challenge. For example, a charge set is *Violent Acquisition* which includes the charges *Robbery*, *Seizure*, and *Kidnapping*. To analyze the effect of causal relationship captured by GCI, they implement a variant which is built upon a correlation-based graph rather than the discovered causal graph. GCI outperforms

Fig. 9.8 An example of part of the generated PAG of the charge *Kidnapping*, as part of the charge set *Violent Acquisition*. *Latent1* and *Latent2* are unobserved confounders. *Latent1* may denote the suspect's intent to commit a crime; and *Latent2* may denote the suspect's intent to obtain properties from the hostage. Taken from [9]

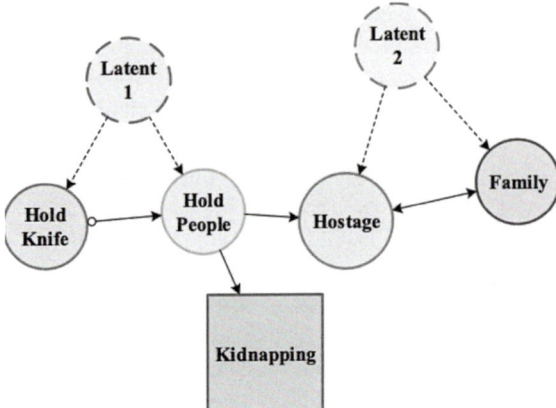

correlation-based method by 4.5% on average accuracy, and 9.8% on average F1, indicating the graph constructed by discovering causal relations better captures the relationship between charges and factors.

Figure 9.8 depicts part of a PAG generated by the system, for the charge *Kidnapping*, as part of the charge set *Violent Acquisition*.

9.6 Learning Causal Structures for Marketing Mix Modeling

How to allocate advertising budget on different channels is a critical issue in marketing. Marketing mix modeling (MMM) addresses this issue as a prediction problem. Ideally, causal MMM should be able to dynamically discover the causal structures of different channels and predict the gross merchandise volume (GMV). However, this is not easy due to two main reasons:

1. Causal Heterogeneity. The causal structures and dynamics of channels vary according to the types of shops and the different time (season) periods.
2. Marketing Response Patterns. The influence of advertising investment has a time decay and could be saturated with more investment.

It is also affected by contextual variables such as economy and competition.

To address these challenges, [6] propose a novel marketing mix model, called CausalMMM. CausalMMM is a graph variational autoencoder-based method, which consists of two main modules: the causal relational encoder and the marketing response decoder. Based on a marketing mix data set, it builds a model that simultaneously (i) infers causal structure among marketing variables for each shop; (ii) predicts the marketing target that satisfies the known marketing response patterns.

The marketing mix data set consists of marketing records from N shops. It is a time series representing the advertising spends on each channel, the marketing target

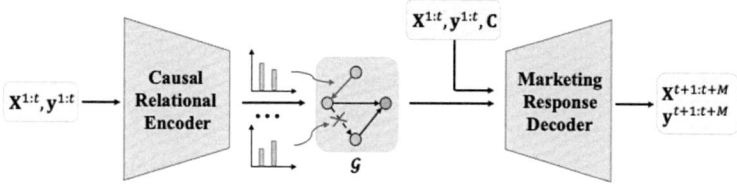

Fig. 9.9 An overview of causalMMM. Taken from [6]

value, and the contextual variables, for each shop. Therefore, the goal of causalMMM is: given a marketing dataset D, infer the causal structures $G_1, ..., G_N$ and predict the marketing target, $Y_t, t = T + 1 : T + M$, for each shop.

An overview of causalMMM is depicted in Fig. 9.9. The causal relational encoder predicts causal structures between marketing variables. The marketing response decoder learns to predict marketing variables given their past observations and contextual variables.

The causal relational encoder aims to infer the likelihood of the causal relations based on the historical advertising spend and marketing target, starting from a fully connected graph. A graph neural network (GNN) is used to propagate information across the fully connected graph and predict causal edges. The edge representation in the graph is obtained through pairwise embeddings, capturing local information in a pairwise manner. Gumbel softmax sampling is applied to derive the structure distribution, obtaining as output the inferred causal structures.

The aim of the marketing response decoder is to model marketing response under the inferred causal structures. It consists of two procedures: the temporal marketing response module, and the saturation marketing response module. To incorporate temporal patterns, it adds sequence models to the original message–passing mechanisms in GNNs. The diminishing returns is very important to marketing decision-making; the saturation marketing response module focuses on S-curve (Hill) transformation, which is the most widely used for modeling saturation. Based on the above two modules, it predicts the value for the advertising channels and the marketing target. For multiple-step forecasting, the prediction results are utilized in a recursive manner.

Both networks, the causal relational encoder and the marketing response decoder, are learned simultaneously through variational inference, where the loss function is composed of two terms: (i) a data-fitting term, and (ii) a structural regularization term.

The performance of CausalMMM was evaluated on two data sets. The first one, is a synthetic data set created to test CausalMMM's ability to reconstruct the causal structures. The second one, is a real-world marketing mix data set collected from an E-commerce platform, which includes 50 shops' marketing mix data from a period of 22 months. Results on the synthetic data set show a high accuracy in the recovery of the causal structures, above 90% in most experiments varying the number of channels and time series length. The results for the real-world data set show competitive performance in terms of prediction error, which varies according to the time horizon; with errors of less then 1% for one time step, approx. 2% for 7 time steps, and approx. 10% for 30 time steps.

9.7 Additional Readings

The proceedings of the 1st and 2nd Workshops on Causal Discovery (CaDis) include several examples of applications of causal discovery in different domains [3,4]. Hassanzadeh et al. describe a demonstration of causal knowledge extraction for large corpuses of text documents [7]. Causal discovery in social media is presented in [11]. Imbens [8] describes methodological approaches for causal inference in statistics and econometrics.

Acknowledgements This chapter is based on the work of different groups as referenced in the corresponding sections. Figures 9.1, 9.2, 9.3, 9.4, 9.5, 9.6, and 9.9 are used with permission. Figures 9.7 and 9.8 are used on the terms of the Creative Commons 4.0 by (Attribution) license: https:// creativecommons.org/licenses/by/4.0/.

References

1. Aguayo-Mendoza A, Irizar-Arrieta A, Casado-Mansilla D, Borges CE (2024) Understanding the factors that affect households' investment decisions required by the energy transition. Plos One 19(3)
2. Aguayo-Mendoza A (2024) Causal modelling to quantitatively analyse people's everyday decisions regarding energy consumption and their reactions to interventions. Doctoral thesis, University of Deusto
3. Sucar LE, Munoz-Benitez JC (eds) (2023) Proceedings of the 1st workshop on causal discovery (CaDis). Academia Mexicana de Computación
4. Sucar LE, Munoz-Benitez JC (eds) Proceedings of the 2nd workshop on causal discovery (CaDis). Academia Mexicana de Computación
5. Campos R, Mangaravite V, Pasquali A, Jorge A, Nunes C, Jatowt A (2020) Yake! keyword extraction from single documents using multiple local features. Inf Sci 509:257–289
6. Gong C, Chen S, Yao D, Li W, Zhang L, Su Y, Bi J (2024) CausalMMM: learning causal structure for marketing mix modeling. In: ACM WSDM'24. Merida, Mexico
7. Hassanzadeh O, Bhattacharjya D, Feblowitz M, Srinivas K, Perrone M, Sohrabi S, Katz M (2020) Causal knowledge extraction through large-scale text mining. In: The thirty-fourth AAAI conference on artificial intelligence, pp 13610–13611
8. Imbens GW (2022) Causality in econometrics: choice versus chance. Econometrica 90(6):2541–2566
9. Liu X, Yin D, Feng Y, Wu Y, Zhao D (2021) Everything has a cause: leveraging causal inference in legal text analysis. In: Proceedings of the 2021 conference of the North American chapter of the association for computational linguistics: human language technologies
10. Ogarrio JM, Spirtes P, Ramsey J (2026) A hybrid causal search algorithm for latent variable models. In: Conference on probabilistic graphical models, pp 368–379

11. Oktay H, Taylor BJ, Jensen DD (2010) Causal discovery in social media using quasi-experimental designs. In: 1st workshop on social media analytics (SOMA). ACM
12. Tang R (2024) Trading with time series causal discovery: an empirical study. arXiv:2408.15846v1
13. Vonk M (2024) Combining literature with causal discovery in environmental-conflict. In: Sucar LE, Munoz-Benitez JC (eds) Proceedings 2nd workshop on causal discovery. Academia Mexicana de Computación
14. Zhang S, Guo B, Dong A, He J, Xu Z, Chen SX (2017) Cautionary tales on air-quality improvement in Beijing. Proc R Soc A: Math Phys Eng Sci 473(2205)

Applications in Artificial Intelligence and Robotics

10

Abstract

In this chapter we describe several works that explore the relation between causal discovery and artificial intelligence. In one case study, a causal Bayesian network is used to reduce the problem of bias when training a machine learning system. A second example, shows how to understand causal relations between events in textual descriptions using large language models. Then we present the incorporation of causal models in generative adversarial networks, providing a finer control of the images generated. The next work, describes how the causal relations between objects in an image can be estimated from the causal direction between pairs of image features based on a neural network classifier. Finally, another case study, shows the use of causal models for accelerating learning in drone navigation, including learning the causal model at the same time that the drone is learning a policy using deep reinforcement learning.

10.1 Introduction

Causal reasoning and artificial intelligence (AI) are interrelated, and each one can benefit the other. Causal models can help to develop ethical AI applications, improve the explanation capabilities of AI systems, increase the learning efficiency and trans-

© The Author(s), under exclusive license to Springer Nature Switzerland AG 2026   203
L. E. Sucar, *Causal Discovery*, Computer Science Foundations and Applied Logic,
https://doi.org/10.1007/978-3-031-98345-0_10

ferability of machine learning, and develop more robust and flexible intelligent systems. AI, in particular machine learning and large language models, could aid causal discovery, in particular from complex and unstructured data such as in computer vision and text processing.

In this chapter we will analyze, through specific examples, several of these relations between causality and artificial intelligence, including the combination of reinforcement learning and causal discovery in robotics; in particular:

- The use of causal models for improving transparency is illustrated in the area of machine learning by identifying possible bias in an hypothetical college admission system.
- The understanding of causal relations between events in textual descriptions using large language models.
- The incorporation of causal models in generative adversarial networks, which provides a finer control of the images generated according to specific characteristics.
- Finding causal relations between objects in an image, by determining the causal direction between pairs of image features based on a neural network classifier.
- How causal models can help robots to learn faster for visual drone navigation, including learning a causal model simultaneously while learning an optimal policy.

10.2   Characterizing Patterns of Unfairness

Data used to train machine learning systems may contain different types of bias (social, cognitive) that can lead to bad decisions. A way to reduce this problem is to analyze the relations between sensitive attributes and the system's outputs; which is a challenging task. DeepMind [7] proposed the use of causal Bayesian networks as a tool to formalize and measure potential unfairness scenarios underlying a data set[1]

The main variables and their relationships in a machine learning system, and in particular those that are sensitive to bias, are modeled as a causal Bayesian network. For example, consider a hypothetical college admission system in which applicants are admitted, A, based on qualifications, Q, choice of department, D, and gender, G; and in which female applicants apply more often to some departments. Figure 10.1

Fig. 10.1 Causal Bayesian
network for the college
admission process

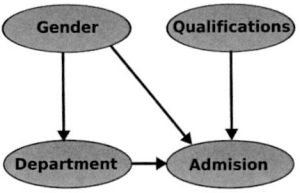

---

[1] This example is based on the information in https://deepmind.com/blog/article/Causal_Bayesian_
Networks..

depicts a CBN of the admission process. Gender has a direct influence on admission through the causal path $G \rightarrow A$ and an indirect influence through the causal path $G \rightarrow D \rightarrow A$. The direct influence represent the fact that applicants with the same qualifications applying to the same department might be treated differently based on their gender. The indirect influence captures differing admission rates between female and male applicants due to their different department choices. The direct influence of the sensitive attribute (gender) on admission is considered unfair for social and legal reasons, the indirect influence could be considered fair or unfair depending on other factors.

CBNs can also be used to quantify unfairness in a dataset and to design techniques for alleviating unfairness in the case of complex relationships in the data. Contrafactual techniques can estimate the influence that a sensitive attribute has on other variables along specific sets of causal paths. This can be used to measure the degree of unfairness on a given dataset in complex scenarios in which some causal paths are considered unfair while other causal paths are considered fair. In the college admission example in which the path $G \rightarrow D \rightarrow A$ is considered fair, path-specific techniques would enable the measurement of the influence of G on A restricted to the direct path $G \rightarrow A$ over the whole population, in order to obtain an estimate of the degree of unfairness contained in the data set.

The use of counterfactual inference techniques would also make possible to ask if a specific individual was treated unfairly. For example, by asking whether a rejected female applicant would have obtained the same decision in a counterfactual world in which her gender were male.

In addition to answering questions of fairness in a data set, path-specific counterfactual inference could be used to design methods to alleviate the unfairness of machine learning systems [6].

10.3 Causal-BERT

Understanding causal relations between events is very important for common sense language understanding and causal discovery. However, finding causal relations between events in texts is a complex task, because it depends on interactions between the particular linguistic expressions, the semantic context established in the text, the specific knowledge of the causal relationships for the domain, and the communicative goals of the text's author. Causality in the text is expressed in arbitrarily complex ways; even when the events are in a single sentence, it might be mentioned in a sparse, ambiguous, or implicit manner. Even explicit causal expressions have rich variation across their syntax, and it is challenging to extract the relevant events and determine causality.

Causal-BERT [12] focuses on causality between events in the same sentence. For this, they used three data sets to train and evaluate their model. SemEval 2007, SemEval 2010, and ADE (Adverse Drug Effects) are publicly available data sets and have annotated events and *Cause-Effect* interactions between them. SemEval 2007 and 2010 are similar, they are both designed to provide a framework for comparing different approaches for classifying different semantic relations in a sentence, including cause–effect relations. The ADE data set is a collection of biomedical texts annotated with drugs and their adverse effects. The authors curated this data set, so it includes the drug names and its effects in each sentence.

The causality-understanding approach can be simplified as a binary classification between *Cause-Effect* vs. *Other* relations between events expressed in natural language. Formally, the problem can be defined as follows: for a given sentence and the two events e_1, e_2, the goal is to predict the possible casual interaction c between the events, where c can be either *Cause-Effect* or *Other*.

To accomplish this task, Causal-BERT is based on BERT, a pre-trained large language model based on a Transformer architecture. BERT provides deep bidirectional representations from the unlabeled text by jointly conditioning on the left and right contexts. BERT based pre-trained models have shown to be very effective at many tasks, including question-answering, relation extraction, etc.

The proposed method involves combining BERT with the three data sets with causal relations to train a binary classifier that can identify cause-effect relations in a sentence. For this, the authors developed three network architectures:

- C-BERT: This is a feed-forward neural network built on top of BERT, which is trained for binary classification of the *Cause-Effect/Other* relations between two events in an input sentence.
- Event-aware C-BERT: This architecture learns a representation of the given sentence expression and predicts causality between the events. Events can be more than a single token, resulting in many vectors when the input sentence is input into a pre-trained BERT model. These are averaged to get the final context of each event expression, and the sentence context, as well as both events' contexts that are fed to a non-linear activation layer followed by a fully connected layer.
- Masked-Event C-BERT: This network architecture is similar to the event-aware C-BERT architecture, where the whole span of event text is replaced with a "BLANK" token. As each event is just a single blank token, unlike Event-aware C-BERT, it docs need to take an average to get the final context of any event. Each model trained by this approach is then fine-tuned using the actual event information provided by the Event-Aware C-BERT model.

10.3.1   Experimental Results

To evaluate the model capabilities in learning the Cause-Effect relations between events, they trained the three different BERT-based network architectures on each of the data sets. The results in terms of F1 score for the three architectures on the three data sets are summarized in Table 10.1.

The results in identifying cause-effect relations in a sentence are very good for the basic architecture, C-BERT, and the other architectures have slightly better results.

Table 10.1 F1 Score for the different architectures on the SemEval 2007, SemEval 2010, and ADE datasets

	SemEval 2007	SemEval 2010	ADE
C-BERT	93.78	97.68	97.10
Event Aware C-BERT	94.94	98.35	97.85
Masked Event C-BERT	95.31	97.85	97.85

Results are similar for the three data sets; with a small decrease in performance for SemEval 2007, probably because it is the smallest of the three data sets.

This work shows that it is possible to extract causal relations within a sentence and in specific contexts; but the problem becomes much more challenging if we want to discover causal relations in texts in a wider context and across several sentences.

10.4 Causal Generative Models

Generative adversarial networks (GANs) are neural generative models that can be trained to imitate sampling from high dimensional nonparametric distributions [9]. A generator network models the sampling process through feedforward computation. The generator output is constrained and refined through the feedback provided by a competitive *adversarial network*, that attempts to discriminate between the generated and real samples. The objective of the generator is to maximize the loss of the discriminator (i.e., convince the discriminator that it outputs images from the real data distribution). GANs have shown great success in generating samples from images and videos.

We would like to generate images according to certain labels, even if these are not in the training images, for instance "women with mustache". For this, we can think of generating an image conditioned on labels as a causal process: *labels determine the image distribution*; so the generator is a functional map from labels to image distributions. The causal graph can also include relations between labels, as labels are not independent and have clear causal connections (e.g., gender causes mustache).

CausalGAN [13] is a framework for conditional image generation based on representing the causal effect between labels and images. They propose causal implicit generative models, which can sample not only from probability distributions but also from conditional and interventional distributions. Based on a causal model, the generator neural networks can be configured to reflect the causal graph structure. For example, the generator neural network architecture depicted in Fig. 10.2, represents the causal Bayesian network: $X \rightarrow Z \leftarrow Y$.

Learning a causal implicit generative causal model is divided into two subtasks. First, learn the causal generative model over a small set of variables. Then, learn the remaining set of variables conditioned on the first set of variables using a conditional

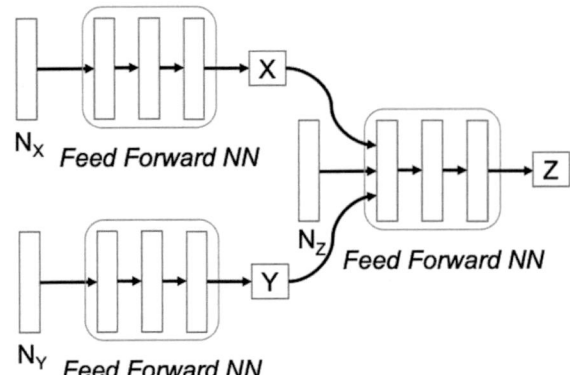

Fig. 10.2 Generator neural network architecture that represents the causal Bayesian network: $X \rightarrow Z \leftarrow Y$. Figure based on [13]

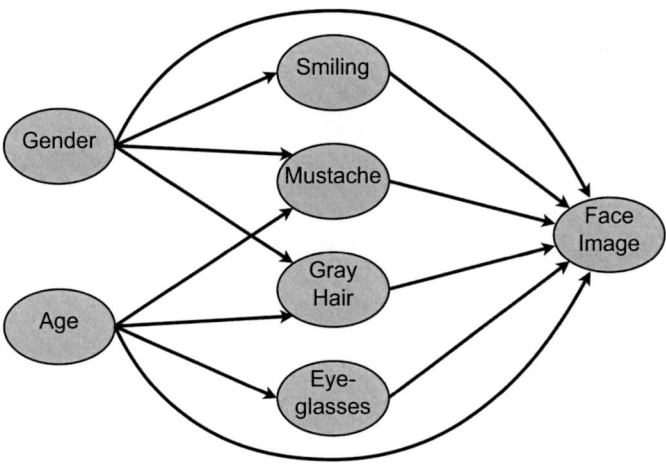

Fig. 10.3 A causal network that represents the causal relations between several characteristics of a face image

generative network. For this to be consistent with the causal structure, every node in the first set should come before any node in the second set with respect to the partial order of the causal graph. For image generation, it first trains a model over the labels, and then it trains a generative model for the images conditioned on the labels. Figure 10.3 shows a possible causal model for generating face images.

10.4.1 Architecture

A novel architecture that can sample from discrete label distributions is proposed [13]. A block diagram of the architecture is depicted in Fig. 10.4. It has two labeler

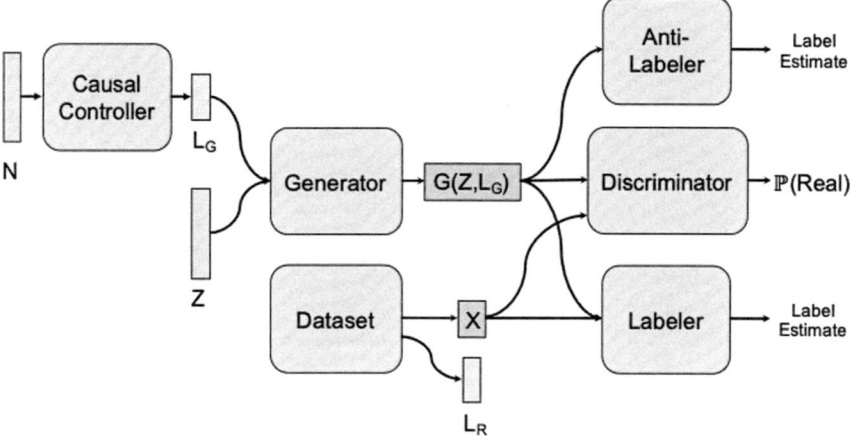

Fig. 10.4 CausalGAN architecture. Figure based on [13]

neural networks. The *Labeler* is trained to estimate the labels of images in the dataset, these labels are generated by the *Causal Controller*. The *Anti-Labeler* is trained to estimate the labels of the images which are sampled from the generator.

The objectives of the generator are: (i) producing realistic images by competing with the discriminator, (ii) capturing the labels it is given in the corresponding images by minimizing the Labeler loss, and (iii) avoiding drifting towards unrealistic image distributions. The generator loss function contains label loss terms, the GAN loss, and an added loss term due to the discriminator. With the incorporation of this additional term to the generator loss, the generator outputs the class conditional image distribution (see [13] for details of the loss function and theoretical guarantees).

The implementation of CausalGAN is based on DCGAN, a convolutional neural net-based implementation of a generative adversarial network; it is modified by adding the Labeler networks, training a Causal Controller network and altering the loss functions. Initially the system is implemented for a single label, and then extended for multiple labels.

10.4.2   Results

CausalGan was tested on the generation of face images, and compared with a conditional GAN. The conditional GAN was implemented based on rejection sampling. This comparison shows the difference between *intervening*, as done with the Causal-Gan, versus *conditioning*, as performed with the conditional GAN.

Figure 10.5 shows several faces generated with CasualGAN (top row), and those generated with the conditional GAN (bottom row), when intervening and conditioning, respectively, on the *Mustache* label according to a causal graph (similar to that in Fig. 10.3).

Fig. 10.5 Face images generated with CausalGAN (top row) by intervening the variable *Mustache* on the causal graph; and face images generated by a conditional GAN (bottom row) by conditioning on *Mustache*. Notice that some images on the top row are females, while all in the bottom row are males. Figure based on [13]

The difference between intervening and conditioning can be observed in Fig. 10.5. Since *Gender* $\rightarrow$ *Mustache* in the causal graph, it is not expected that $do(Mustache = 1)$ affects the probability of *Gender*. The top row, with images generated by the CausalGAN, shows both males and females with mustaches, even though the generator never sees the label combination *Gender* = *Male* and *Mustache* = 1 during training. The bottom row, obtained from the conditional GAN, only shows images of men with mustaches, as in the dataset $P(Gender = Male|Mustache = 1) = 1$.

The previous difference between the CausalGAN and the conditional GAN is clear due to the way in which causal reasoning is performed. When we intervene on *Mustache*, the link pointing to it from *Gender* is deleted, so *Gender* does not affect the results of this intervention. However, when we condition, the link is not removed, and *Gender* influences the results.

10.5 Causal Discovery from Images

Most work on causal discovery from images requieres the definition of abstract features obtained from the images, which are usually defined by a person (as we will discuss for drone navigation, in the next section). A more challenging task is causal discovery directly from the images, without requiring the manual definition of abstract features.

A deeper understanding of images requires the ability to reason about how the scene depicted in the image would change in response to *interventions*. For example, given an image of a car circulating in a bridge over a river, a possible question could be: *How would the scene look like if we remove the bridge?* [14]. If the bridge had been removed from the scene, it would make no sense to observe the car floating weightlessly above the river. Thus, the presence of the bridge has an effect on the presence of the car; that is, the presence of the bridge *causes* the presence of the car, in the sense that if the bridge was not there, then the car would not be either.

The previous problem is related to the *dispositional semantics of causality*, in which causal relations are established when objects exercise some of their causal dispositions or the *powers of objects*. For example, a bridge has the power to provide support for a car, and a car has the power to cross a bridge. In this context, [14] address the following question: Can asymmetric causal relationship be inferred from statistics observed in image datasets? We say that the objects of category A cause the presence of objects of category B when $C(A, B)$ is greater than the opposite, $C(B, A)$. This induces a network of asymmetric causal relationships between object categories that represents how real-world scenes would be modified when one were to make certain objects disappear. Answering this question will provide a way for computer vision algorithms to reason about the causal structure of the world.

As no data is available to respond the previous question, the authors [14] propose an indirect method based on distinguishing between object features and context features, for different classes of objects. Object features correspond to those inside the bounding box of the object (provided by object detection algorithm), context features are those outside the bounding box. Additionally, they distinguish between causal features and anti-causal features. Causal features are those that cause the presence of the object in the scene, whereas anti-causal features are those caused by the presence of the object in the scene. Based on these definitions, they make the following hypothesis: "There exists an observable statistical dependence between object features and anti-causal features. The statistical dependence between context features and causal features is nonexistent or much weaker". Intuitively, we expect that many of the features caused by the presence of certain object are parts of the object itself, so they are likely to be contained inside its bounding box. In contrast, the context of an object of interest may either cause or be caused by the presence of the object.

To find out the causal relations between object features and context features in images, [14] developed a neural network classifier. Although in general it is impossible to find out the direction of causality between two variables, in some cases it can be determined (i.e., linear non-Gaussian models such as LINGAM, see Sect. 4.3.5.1). In such cases, the causal direction between the variables of interest may leave a detectable signature in their joint distribution. They exploit this insight to build a classifier for determining the cause-effect relation between two random variables from samples of their joint distribution. The causality classifier was trained on synthetic cause-effect data and tested on real data [16] with good results, it correctly classifies the cause-effect direction of 79% of observational samples in the real data.

To test their hypothesis, they consider twenty object categories of the Pascal VOC 2012 dataset [8], using a feature extraction network trained on ImageNet and a classifier network trained on the Pascal dataset. Then, they use these networks to identify causal relationships on the subset of the 99,309 MSCOCO images representing objects from twenty Pascal categories: airplane, bicycle, bird, boat, bottle,

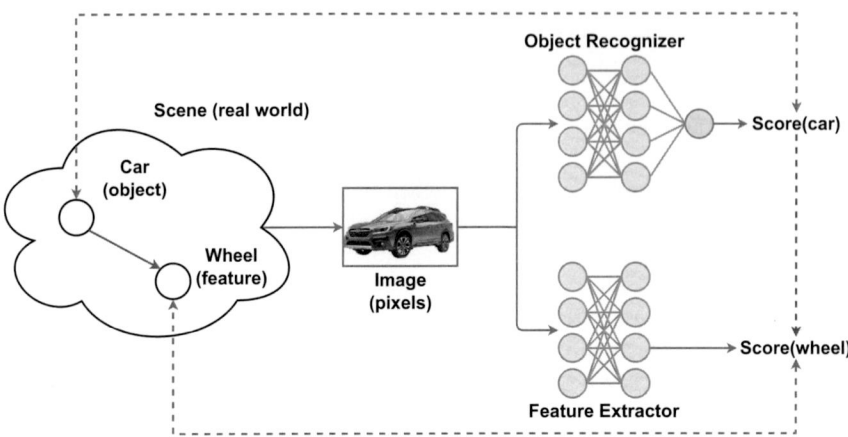

Fig. 10.6 The figure illustrates the proposed method in which, based on a pretrained causal classifier, the causal relations between the feature variables (obtained by the feature extractor) are computed from the image pixels. The appearance of causation between the image features of the real world entities, in this example the car and the wheel (detected by the object recognizer), suggests that there is a causal link between the real world entities themselves: *the presence of the car causes the presence of a wheel*. Figure based on [14]

bus, car, cat, chair, cow, dining table, dog, horse, motorbike, person, potted plant, sheep, sofa, train, and television.

As features they use the output vector of the last layer of the feature extraction neural network, composed of 512 real values obtained from the 20 object categories. For each object category and each feature, they apply the causal classifier to the scatter plot representing the joint distribution of the scores of feature j and the score of category k. Since these scores are computed by running the neural networks on the image pixels, they are not related by a direct causal relationship. However, these scores are highly correlated with the presence of objects and features in the real scene. Therefore, the appearance of a causal relation between these scores suggests that there is a causal relationship between the real world entities they represent. For example, if a causal relation is obtained between features of a car and those of a wheel, we can hypothesize that *the presence of the car causes the presence of a wheel*, see Fig. 10.6.

In order to verify their hypothesis, it is sufficient to show that the top anti-causal features are more likely to be object features than the top causal features. For each category k and each feature j, we must determine whether feature j is likely to be an object feature or a context feature. This can be done by blacking out pixels in the image that are inside the object's bounding box (object-features) or those outside the bounding box (context-features), and in this way distinguish between causal and anti-causal features. The experiments obtained for the 20 object categories show that the object features are related to anti-causal features: the top 1% anti-causal features have a higher object-feature ratio than the top 1% causal features, therefore providing strong support for the hypothesis. Also, this indicates that anti-causal features may be useful for detecting objects locations in a robust manner, regardless of their context.

Although this work provides some evidence that it is possible to detect causal relations between image features, there are still many challenges so that computer vision algorithms will be able to perceive the causal structure of the real world and reason about scenes.

10.6   Visual Drone Navigation Incorporating Causal Models

The use of Unmanned Aerial Vehicles (UAV) or drones has increased in recent years, due to the interest in using them for entertainment, but also for military, agriculture, delivery services, and rescue applications, among others. A limitation is the need for a human pilot to control it, so multiple research works have used different artificial intelligence techniques to develop autonomous pilots. One of the most popular techniques is Deep Reinforcement Learning (DRL) [1], which has been used for collision avoidance and seeking an objective in different environments.

A challenge for DRL in robot navigation is the large state space, so it takes a long time to learn an optimal policy, which makes it impractical even in simulation. Incorporating knowledge from a causal model into DRL algorithms, can greatly speed up the learning process by reducing the need for extensive exploration.

Xolo-Tlapanco et al. [19] have developed a DRL algorithm that, in a first stage, incorporates a causal model to accelerate learning the policy. In a second stage, they do not provide the causal model *a priori*, but this is learned in the process of learning the optimal policy. They use Deep Q-Learning (DQN) [15] as the DRL algorithm to implement the visual navigation task in the drone. The original algorithm is modified to incorporate the causal knowledge represented as a causal Bayesian network (CBN), by modifying the action selection policy during training.

If the CBN model is already known, it can be used from the onset of the learning process. At each step, it takes the state elements present in the Markov blanket of the reward node as evidence, which is used in the inference of reaching each state value. This is divided in two parts, one for the probability of transition to a positive reward and other for the probability of transition to a negative reward. If the selected action has a high probability to obtain a positive reward, the action is executed. Otherwise, if the action selected has a high probability to reach a negative reward, the action is eliminated as an option and another one is randomly selected. If none of the previous conditions is satisfied, the algorithm applies the standard exploration policy, i.e., ϵ-greedy.

If the CBN is not provided, the drone should learn it in the process of learning the policy using DRL. For this, it stores at each step, the current state, the action executed, the next state, and the reward obtained; that is the tuple (S_t, a_t, S_{t+1}, r_t). Every k steps during the learning process, the CBN model is updated based on the stored tuples. A score-based causal discovery algorithm is used to learn the CBN, with a hill-climbing search strategy and the BIC score. The causal network is represented as a two-stage dynamic Bayesian network for each action. In the initial k steps no causal model is available, so it performs the standard exploration process. After k steps a causal model is learned and then used for the next k steps when the causal model is updated, and this process is repeated until the end of the learning stage.

10.6.1   Experimental Evaluation

The previous approach was evaluated in a drone navigation task in which it has to reach a particular object based on vision, combining an RGB camera and a depth camera. The configuration of the experiment was the following:

State space: An image with dimensions $84 \times 84 \times 3$ pixels.

Action space: Eight discrete actions—forward and backward moves (move the drone in the y axis), turn left and turn right (change the orientation of the drone in the x axis), right and left (move the drone in the x axis), and ascend and descend (move the drone in the back z axis).

Target: Goal image.

Reward: +100 target recognized, +1000 goal is reached, –1000 drone crashes, –9 otherwise.

For the causal model representation and learning, the state space was transformed into an abstract representation with nine discrete variables: (i) Distance to five defined sections of the image (center, top left, top right, bottom left, and bottom right). In each section, a number of random points (at least 20 points) is selected, taking the smallest distance as the overall distance of the drone to the objects in this section— five variables. (ii) A boolean variable, true if the goal is in the field of view—one variable. (iii) Distance and angle to the goal—two variables. (iv) Altitude of the drone—one variable. All these variables are discretized, taking two values for the distance: close and far; a boolean value for the goal in sight; three values for the angle: center, far left and far right, and three values for the altitude: good, close and far from the ground.

A causal model was defined by an expert in drone navigation for each action, represented as a two-stage dynamic Bayesian network, given a total of eight graphs. These causal graphs include the relevant abstract state variables at times t and $t + 1$, and the reward at time $t + 1$. Three of these networks are shown in Fig. 10.7.

To obtain the information for the abstract state representation, three sensors are required: an RGB camera to process the image to detect the goal, a Depth Camera to obtain the distances, and a barometer or similar to obtain the altitude of the drone. The goal is a sign of *Heliport*, which is detected in the image using YOLO v8 [17], trained to recognize this sign.

In an initial experiment, the impact of the number of steps between updating the causal model was evaluated, with several values of k: 100, 200, 400, and 600. $k = 200$ was selected as it tends to provide more consistent and stable results.

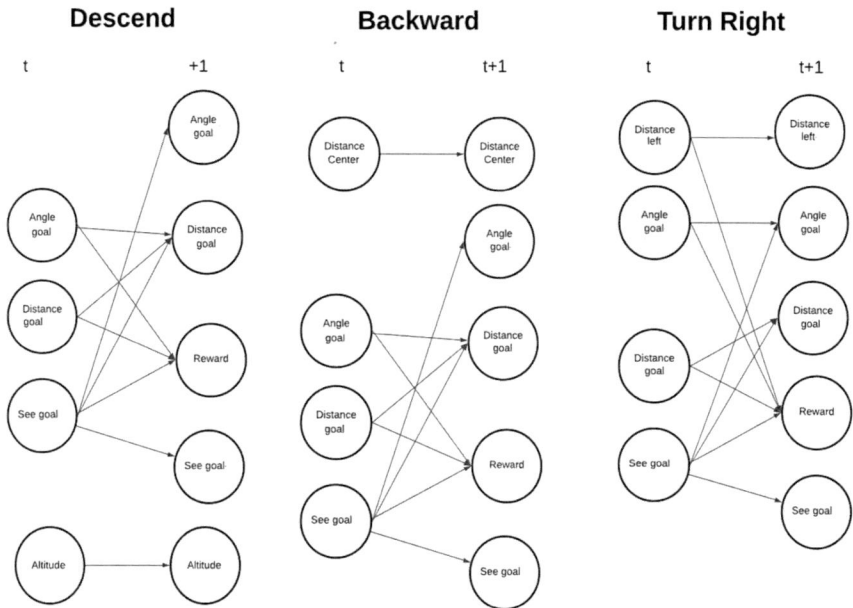

Fig. 10.7 CBNs defined by an expert relating abstract state variables at consecutive time steps, t and $t + 1$, and reward at $t + 1$, for the actions *descend, backward*, and *turn right*

Then, the following variants of the DRL algorithms were compared (i) DQN without causal knowledge as the baseline, (ii) DQN incorporating the *expert* defined causal model from the start, (iii) three versions of the method that learns the causal model, all with $k = 200$; one that considers only positive rewards (CBN-DQN P), one that considers positive and negative rewards (CBN-DQN PN), and an improved version (CBN-DQN v2) that consults the causal model *before* executing a random action according to the exploration strategy.

The performance of the five algorithms during training is illustrated in Fig. 10.8, depicting in the vertical axis the reward obtained in each episode after applying a moving average of size 50, and in the horizontal axis the number of training episodes. The best behavior is observed when the Causal Bayes Network is manually defined. The worst with the basic DQN without causal knowledge. The performance of the three variants of the method that learns and uses the causal model are similar, with slight advantage for CBN-DQN v2. It seems that the variants learning the causal model can obtain a similar performance to the one with the prior model if given more training episodes.

The policy, learned in simulation, was used with a real drone with satisfactory results in a laboratory environment. An image from one of the tests while the drone is navigating autonomously to the goal (H) is shown in Fig. 10.9.

The incorporation of causal knowledge can accelerate learning in the context of DRL, as shown in this application. However, a limitation is that it requires the manual

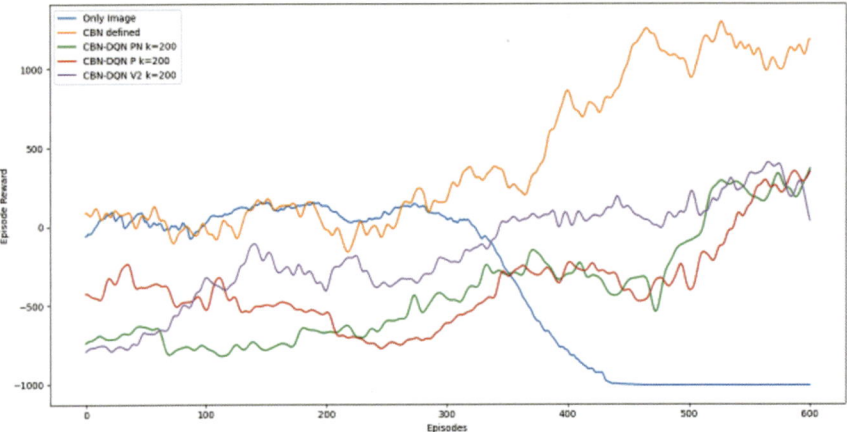

Fig. 10.8 Performance of several DRL algorithms in the task of controlling a drone to approach a goal, including: (i) DQN without causal knowledge, (ii) DQN incorporating prior causal knowledge, (iii) Three variants of the algorithm that learns and uses the causal model. Figure from [19]

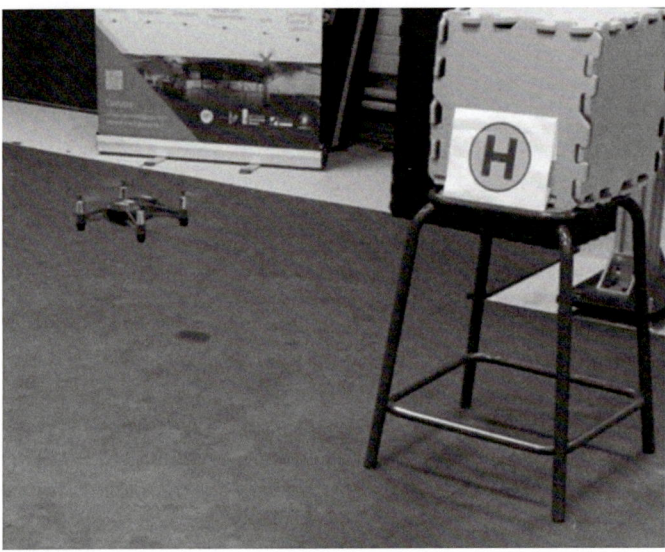

Fig. 10.9 An image of a drone while navigating to the goal (**H** sign) using the policy learned in simulation aided with a causal model

specification of relevant abstract variables that summarize the state space, a challenge is to learn automatically this abstract representation useful for causal discovery.

10.7 Additional Reading

Cavique [5] presents a review of the implications of causality in artificial intelligence. The role of causality in explainable artificial intelligence is investigated in [2]. Castri et al. [3] propose efficient causal discovery for robotic applications. A practical focus on causal inference is presented in [11], including several examples of applications. An analysis and guidelines for the use of causality in medical image analysis is described in [4]. A method to generate new images based on counterfactual estimation from observational imaging is presented in [18]. A survey on the integration of causal inference and deep learning is given in [10].

Acknowledgements The different applications presented in this chapter are based on the work of the referenced authors. Figures 10.2, 10.4, 10.5, and 10.8 are used with permission.

References

1. Aggarwal CC (2023) Neural networks and deep learning. Springer Nature Switzerland
2. Carloni G, Bertia A, Colantonio S (2023) The role of causality in explainable artificial intelligence. arXiv: 2309.09901v1
3. Castri L, Mghames S, Bellotto N (2023) Efficient causal discovery for robotics applications. arXiv:2310.14925v2
4. Castro DC, Walker I, Glocker B (2020) Causality matters in medical imaging. Nature Commun 22;11(1):3673
5. Cavique L (2024) Implications of causality in artificial intelligence. Front Artif Intell 7:1439702
6. Chiappa S, Gillam TPS (2018) Path-specific counterfactual fairness. arXiv-stat.ML
7. Chiappa S, William SI (2019) A causal bayesian networks viewpoint on fairness. In: IFIP advances in information and communication technology, pp 3–20
8. Everingham M, VanGool L, Williams CKI, Winn J, Zisserman A (2012) The PASCAL visual object classes challenge 2012 (VOC2012)
9. Goodfellow I, Pouget-Abadie J, Mirza M, Xu B, Warde-Farley D, Ozair S, Courville A, Bengio Y (2014) Generative adversarial nets. In: Proceedings of neural information processing. Montreal, Canada
10. Jiao L, Wang Y, Lui X, Li L, Lui F, Ma W, Guo Y, Chen, P, Yang S, Hou B (2024) Causal inference meets deep learning: a comprehensive survey. Res Jl 7, Article 0467
11. Kamath U, Graham K, Naylor M (2023) Applied causal inference. Amazon, France
12. Khetan V, Ramnani R, Anand M, Sengupta S, Fano A (2021) Causal-BERT : language models for causality detection between events expressed in text. In: Intelligent computing: proceedings of the 2021 computing conference, vol 1. Springer International Publishing, pp 965–980
13. Kocaoglu, M., Snyder, C., Dimakis, A., and Vishwanath, S.: CausalGAN: Learning Causal Implicit Generative Models with Adversarial Training. Proceedings International Conference on Learning Representations (2018)

14. Lopez-Paz D, Nishihara R, Chintala S, Scholkopf B, Bottou L (2017) Discovering causal signals in images. In: Proceedings of computer vision and pattern recognition conference
15. Mnih V, Kavukcuoglu K, Silver D, Rusu, AA Veness J, Bellemare MG, Graves A, Riedmiller M, Fidjeland AK, Ostro-vski G, Petersen S, Beattie C, Sadik A, Antonoglou I, King H, Kumaran D, Wierstra D, Legg S, Hassabis D (2015) Human-level control through deep reinforcement learning. Nature 518(7540):529–533
16. Mooij J, Peters J, Janzing D, Zscheischler J, Scholkopf B (2016) Distinguishing cause from effect using observational data: methods and benchmarks. J Mach Learn Res
17. Redmon J, Divvala S, Girshick R, Farhadi A (2016) You only look once: Unified, real-time object detection. In: IEEE conference on computer vision and pattern recognition (CVPR), pp 779–788
18. Sanchez P, Tsaftaris SA (2022) Diffusion causal models for counterfactual estimation. In: 1st conference on causal learning and reasoning, Proceedings of machine learning research, vol 140, pp 1–21
19. Xolo-Tlapanco N, Morales E, Sucar LE, Gomez-Balderas E (2024) Visual robot navigation incorporating causal models in deep reinforcement learning. In: 2nd workshop on causal discovery (CaDis). IBERAMIA, Montevideo, Uruguay

Causal Discovery Tools

There is a wide variety of software tools for causal reasoning and causal discovery, here we list some of them (in alphabetical order), including information on where to find them.

Bambi: A tool for causal (mixed-effects) modeling in Python. It works with the PyMC probabilistic programming framework: https://github.com/bambinos/bambi

Causal-learn: A Python library for causal discovery that provides a translation and extension of the Tetrad java code. It offers the implementations of up-to-date causal discovery methods as well as simple and intuitive APIs: https://causal-learn.readthedocs.io/en/latest/

CausalML: A Python package that provides a suite of uplift modeling and causal inference methods using machine learning algorithms. It provides a standard interface that allows user to estimate the Conditional Average Treatment Effect (CATE): https://causalml.readthedocs.io/en/latest/about.html

CausalNex: The CausalNex library enables practitioners to learn structural relationships from data and allow domain experts to verify the accuracy of the relationships between different data sets:
https://medium.com/quantumblack/introducing-causalnex-driving-models-which-respect-cause-and-effect-a561545f0a5e

DoWhy: DoWhy is a Python library for causal inference that supports explicit modeling and testing of causal assumptions. DoWhy is based on a unified language for causal inference, combining causal graphical models and potential outcomes frameworks: https://github.com/py-why/dowhy

TETRAD: A desktop Java application that creates, simulates data from, estimates, tests, predicts with, and searches for causal models:
https://www.ccd.pitt.edu/tools/

Tigramite: A Python package for causal inference with a focus on time series data: https://github.com/jakobrunge/tigramite

© The Editor(s) (if applicable) and The Author(s), under exclusive license to Springer Nature Switzerland AG 2026
L. E. Sucar, *Causal Discovery*, Computer Science Foundations and Applied Logic,
https://doi.org/10.1007/978-3-031-98345-0

Glossary

Association A type of reasoning in which regularities in the world are identified.

Average Treatment Effect (ATE) The average difference in the variable of interest (effect) between the results of both groups in a randomized clinical trial.

Bayesian network A directed cycling graph that represents the joint distribution of a set of random variables such that each variable is conditionally independent of its non-descendants given its parents in the graph.

Causation A stochastic relation between events in a probability space; this is, some event (or events) causes another event to occur. It is transitive, irreflexive and anti-symmetric.

Causal Bayesian Network A directed acyclic graph in which the nodes represent random variables and the arcs correspond to direct causal relations.

Causal discovery The process of learning a causal model from observational and/or interventional data.

Causal Markov Condition Every variable is independent of its nondescendents given its parents in a causal graph.

Causal Faithfulness The probability distribution $P(V)$ over a set of variables V is faithful to the graph G, if all and only the independence relations true in $P(V)$ are entailed by the causal Markov condition.

Causal ordering A causal ordering, $X_1, X_2, ..., X_n$, of a causal graphical model (DAG), implies that there could not be arcs from variables which have a higher number to variables with a lower number according to the order.

Causal prediction A type of reasoning that predicts the effect of deliberative actions.

Causal Reinforcement Learning The integration of causal models and reinforcement learning.

Causal reasoning A procedure for answering causal queries from a causal model.

Causal sufficiency There are no unmeasured cofactors (common causes) of the observed variables in a causal model.

Classifier A method or algorithm that assigns labels to objects.

Clique A completely connected subset of nodes in a graph that is maximal.

© The Editor(s) (if applicable) and The Author(s), under exclusive license to Springer Nature Switzerland AG 2026
L. E. Sucar, *Causal Discovery*, Computer Science Foundations and Applied Logic,
https://doi.org/10.1007/978-3-031-98345-0

Cofactor A variable that influences (causes) the cause and effect variables in a causal model (ancestor of both variables in a graphical model).

Completed Partially Directed Acyclic Graph A hybrid graph that represents a Markov equivalence class of DAGs.

Complete graph A graph that has an edge between every pair of vertices.

Conditional independence Two variables are conditionally independent given a third variable if they become independent when the third variable is known.

Conditional probability Probability of certain event given that another event or events have occurred.

Counterfactuals A type of reasoning that predicts what would have been true under different circumstances under an intervention (What would have happened if?).

Covariate Cofactor.

Deep reinforcement learning Reinforcement learning in large or continuos state spaces, in which function approximators, usually represented as neural networks, are used to approximate the value function or policy.

Directed Acyclic Graph A directed graph that has no directed circuits (a directed circuit is a circuit in which all edges in the sequence follow the directions of the arrows).

Directed Graph A directed graph or digraph is an ordered pair, $G = (V, E)$, where V is a set of vertices or nodes and E is a set of arcs that represent a binary relation on V.

Do-Calculus The Do-Calculus consists of a set of inference rules that allows to transform a probabilistic query about interventions, to other query based on observed quantities (ordinary conditional probabilities).

D-separation A graphical criteria for determining if two subsets of variables are conditionally independent given a third subset of variables in a Bayesian network.

Dynamic Bayesian Network An extension of Bayesian networks to model dynamic processes; it consists of series of time slices, each time slice represents the state of all variables at certain time.

Endogenous Variable A variable in a causal model that is not exogenous.

Event sequence Consists of multiple events E_t which are organized in a specific order based on their time of occurrence.

Exogenous Variable A variable in a causal model that is a root node in the causal graph (it has no parents).

Expectation-Maximization An statistical technique used for parameter estimation when there are non-observable variables.

Expected Utility Average gain of all the possible results of a decision, weighted by their probability.

Experiment An experiment E on a set of variables V is represented by a triplet of sets $E = (S, R, I)$, where S represents the subset of variables in the model that is subject to an intervention, I is the corresponding set of intervention variables, and R contains the remaining passively observed variables.

Functional Causal Model Structural Equation Model.

Gaussian Linear Model A type of Structural Equation Model that consists of a set of linear equation with Gaussian noise.

Granger Causality A definition of causality based on prediction improvement. X causes Y if there is some information in X relevant for Y that is not contained in Y's past.

Graph A graphical representation of binary relations between a set of objects.

Identifiability A causal effect is identifiable if it can be determined from a causal model and the associated observational data.

Imputation Imputation is the process of replacing missing data with substituted values.

Independence Axioms In the context of Bayesian networks, a set of rules to derive new conditional independence relations from other conditional independence relations.

Independent variables Two random variables are independent if knowing the value of one of them does not affect the probability distribution of the other one.

Influence Diagram A graphical model for solving decision problems. It is an extension of Bayesian networks that incorporates decision and utility nodes.

Information In abstract terms it can be though as the resolution of uncertainty. In Information Theory it refers to a measure of what a message communicates, and it is inversely proportional to its probability.

Instrumental variable A variable, Z, (called an instrument) that it is added to a causal model, such that Z is correlated with the cause variable, X, and not correlated with the effect variable, Y, so that the causal effect of X on Y can be estimated.

Interpretation of Probability Meaning of the primitive symbols of the formal theory of probability.

Intervention A structural or parametric intervention.

Joint Probability The probability of a conjunction of N propositions.

Junction Tree A tree in which each node corresponds to a subset (group or cluster) of variables.

Knowledge graph A causal graph that contains different types of prior information for a particular edge, such as: (i) adjacency (connected but do not know the direction), (ii) non-adjacency, (iii) directed cause, (iv) semi-directed cause (non-adjacency or direct cause).

Maximal Ancestral Graphs A hybrid graph that represents conditional independence and causal relationships in DAGs that include unmeasured (hidden or latent) variables.

Marginal Probability The probability of an event independent of information on other events.

Markov Blanket A set of variables that make a variable independent of all other variables in a probabilistic graphical model.

Markov Decision Process A model for sequential decision making composed of a finite set of states and actions, in which the states follow the Markov property.

Markov Equivalence Graphical models that codify the same conditional independence relations.

Markov Property The probability of the next (future) state is independent of the previous states (past) given the current (present) state.

Minumum Description length A measure that makes a compromise between accuracy and model complexity.

Multi-connected Directed Acyclic Graph A directed acyclic graph in which there is more than one path between some pairs of nodes.

Multivariate point process It refers to a stochastic process consisting of a series of binary events that occur in continuous time.

Multivariate time-series It consists of N time series X where, for a certain time t, each X_t is a vector $X_t = (X_{t1}, X_{t2}, ..., X_{tN})$ in which each variable X_{ti} represents a measurement of the i-th time series at time t.

Parametric intervention Parametric interventions are a weaker form of intervention, also known as partial, soft, conditional or dependent intervention, in which the post-manipulation graph is changed by the addition of the intervention variables. Parametric interventions do not alter the structure, instead they modify the parameters of the conditional distribution of the intervened variables.

Partial Ancestral Graph A Markov equivalence class of maximal ancestral graphs.

Policy A function that maps states to actions.

Polytree A singly-connected directed acyclic graph.

Probabilistic Graphical Model A compact representation of a joint probability distribution of a set of random variables composed by a graph and a set of local probability distributions.

Probabilistic Inference A procedure for calculating the posterior probability of the unknown variables in a probabilistic graphical model given certain evidence (a subset of known or instantiated variables).

Probability A function that assigns a real number to each event (subset of a sample space) and satisfies certain axioms known as the probability axioms.

Randomized Control Trial An experimental study in which certain treatment is applied to an *experimental* group of individuals, and alternative treatment (or a placebo) is applied to a *control* group; the members of both groups are selected at random within certain population. The results of both groups in terms of a variable(s) of interest are compared to estimate the causal effect of the treatment.

Random Variable A mapping from a sample space to real numbers.

Rational Agent An agent that selects its decisions to maximize its expected utility according to its preferences.

Reinforcement learning Reinforcement learning is learning what to do, how to map situations to actions so as to maximize a numerical reward signal.

Sample space The set of possible outcomes of an experiment.

Singly-connected Directed Acyclic Graph A directed acyclic graph in which there is a single path between every pair of nodes.

Skeleton An undirected graph that keeps the same adjacencies of the corresponding directed graph (without the directions of the arcs).

Structural Causal Model A causal model that consists of a set of structural equations.

Structural Equation A functional equation that relates a variable, X, to its direct causes that correspond to the parents of X, $Pa(X)$, in a graphical causal model.

Structural Equation Model Structural Causal Model.

Structural intervention Structural interventions make the intervened variable independent of its normal causes; this type of interventions are also referred to as randomizations, surgical interventions, or ideal interventions.

Subsampling Data sampling of a time series for which causal interactions may occur on a time scale faster than the frequency of measurement.

Temporal Event Network A Bayesian network for modeling dynamic processes in which some nodes represent the time of occurrence of an event or state change of certain variable.

Transfer Learning To transfer knowledge and/or data from related domains or tasks to learn another task.

Tree A connected graph that does not have simple circuits.

Undirected Graph An undirected graph is an ordered pair, $G = (V, E)$, where V is a set of vertices or nodes and E is a set of edges that represent symmetric binary relations.

Index

© The Editor(s) (if applicable) and The Author(s), under exclusive license to Springer
Nature Switzerland AG 2026
L. E. Sucar, *Causal Discovery*, Computer Science Foundations and Applied Logic,
https://doi.org/10.1007/978-3-031-98345-0

Zeitfracht Medien GmbH
Ferdinand-Jühlke-Straße 7
99095 Erfurt, Deutschland
produktsicherheit@kolibri360.de